MICHIGAN.

BEING

CONDENSED POPULAR SKETCHES

OF THE

Topography, Climate and Geology

OF THE STATE.

BY
ALEXANDER WINCHELL, LL. D.,
CHANCELLOR OF THE SYRACUSE UNIVERSITY; LATE PROFESSOR OF GEOLOGY, ZOOLOGY AND BOTANY IN THE UNIVERSITY OF MICHIGAN.

[Extracted, by permission, from Walling's Atlas of Michigan.]

PRINTED BY
THE CLAREMONT MANUFACTURING COMPANY.
1873.

NOTE.

The author's attention, during a long residence in Michigan, was directed, both by his natural tastes and by professional duties, to the physical characteristics of the State; and, in this study, he accumulated a large fund of information which has never been published. The following papers are rather hastily prepared and incomplete abstracts of the information in his possession respecting the Topography, Climate and Geology of the State. Arrangements have been made for publishing the complete details of the Topography.

TOPOGRAPHY AND HYDROGRAPHY.

BY ALEXANDER WINCHELL, LL. D.

Professor of Geology, Zoölogy, and Botany in the University of Michigan. Late Director of the State Geological Survey.

THE STATE OF MICHIGAN occupies a position approximating the centre of the continent of North America. The geographical centre of the continent is not far from the Lake of the Woods, which is 560 miles in a straight line from the centre of the State, and 260 miles from its western extremity. The centre of the State is marked by the position of Carp Lake, in Lelanau county, which is 670 miles in a straight line from New York, the nearest point on the Atlantic Seaboard. The State is limited by natural boundaries on all sides except the south. Politically,* it has 708.5 miles coter-

* The political boundaries of the State are defined by the following documents: Sixth Article of the Treaty of Ghent; Report of Commisioners provided by that Article, and dated June 18, 1822; Act admitting Michigan into the Union, June 15, 1856; Act

minous with the Dominion of Canada; 55.5 miles coterminous with Minnesota; 571 miles coterminous with Wisconsin; 58 miles bordering on Illinois; 129.2 miles on Indiana, and 92.8 miles on Ohio; making a total length of boundary line, amounting to 1615 miles.

The land area of the State consists of two natural divisions, known as the Upper and Lower Peninsulas, to which are attached the contiguous islands. The Upper Peninsula is bounded by portions of the lakes Superior, Michigan and Huron, the river St. Mary and the State of Wisconsin. The Lower Peninsula is embraced by lakes Michigan, Huron, St. Clair and Erie, and the St. Clair* and Detroit rivers; and is bounded on the south by the States of Ohio and Indiana.

The main land of the State is embraced between the parallels of 41° 692 and 47° 478 north latitude, and the meridians of 82° 407

of April 19, 1816, Sec. 2; Act admitting Wisconsin, Aug. 6, 1846; Act of April 18, 1818, Sec. 2. For the original boundary of Ohio on the north, see Act of April 30, 1802.

* The river St. Clair was originally named Sinclair from Patrick Sinclair, a British military officer, who purchased of the Indians, in 1765, 4000 acres of land on the river. Lake St. Clair was so named from a French officer. (American State Papers, Public Lands, Vol. I, p. 246.)

and 90° 536 of longitude west from Greenwich. The most northerly point is the north side of Keweenaw Point, five miles west of the Light House at Copper Harbor; and the most southerly is the north-west corner of Ohio. The most easterly point is at Port Huron, near the outlet of Lake Huron; and the most westerly is at the mouth of Montreal river. The most northern territory belonging to the State, is Gull Islet, off the extremity of Ile Royale, which attains the latitude of 48° 211.

The following table exhibits the latitudes and longitudes of the principal points of the State:

TABLE OF GEOGRAPHICAL POSITIONS.

STATIONS.	LATITUDE.			LONGITUDE.		
	°	′	″	°	′	′
Detroit, St. Paul's Church,	42	19	45.85	83	02	22.73
" Congrega'l Church,	42	19	45.64	83	02	29.07
" Intersection Fort and Griswold Sts.,	42	19	49.85	83	02	20.63
Fort Gratiot, Light House,	43	00	21.86	82	24	43.96
Pt. aux Barques, Light House,	44	01	23.35	82	47	09.87
Saginaw, Light House,	43	38	37.84	83	50	54.46
Tawas, Light House,	44	15	35.44	83	26	14.57
Mouth of Thunder Bay River,	45	03	38.90	83	25	32.63
Detour Light House,	45	57	20.11	83	54	21.71
Fort Holmes, Mackinac I.,	45	51	27.81	84	36	24.48
Waugoshance Light House,	45	47	13.38	85	04	56.83
N. E. cor. Big Beaver Island,	45	45	12.67	85	29	38.00
Sand Point, Escanaba,	45	44	35.04	87	02	25.65
Menominee,	45	05	19.31	87	35	25.20

STATIONS.	LATITUDE.			LONGITUDE.		
	°	′	″	°	′	″
Grand Haven, Court House,	43	03	47.25			
" Lake Survey Sta.	43	03	50.14	86	14	21.30
Marquette, Light House,	46	32	55	87	22	12.45
Vulcan, near Copper Harbor,	47	26	44.25			
Ann Arbor, Observatory,	42	16	48.30	83	43	43.05
New Buffalo, Intersection of middle of Whittaker Ave. and Mechanics St.,	41	47	47.00	86	44	53.55
Niles, Steeple of Trinity Church,	41	49	46.10	86	15	36.60
Monroe, Light House,	41	53	26.77	83	19	22.29
Adrian,	41	54	26	83	59	27
Hillsdale,	41	55	19	84	33	46
Coldwater,	41	53	30	85	01	32
White Pigeon,	41	44	59	85	39	42
Ypsilanti,	42	14	12	83	37	06
Jackson,	42	14	46	84	23	01
Marshall,	42	13	38	84	56	09
Kalamazoo,	42	17	39	88	35	58
Allegan,	42	31	49	85	52	37
Lansing,	42	43	53	84	30	42
Pontiac,	42	37	44	83	17	21
Owosso,	43	00	17	84	18	21
Grand Rapids,	42	57	59	85	39	59
Muskegon,	43	13	54	86	15	51
Flint,	43	01	01	83	40	58
Tuscola,	43	19	31	83	39	20
East Saginaw,	43	26	25	83	55	43
Manistee,	44	13	41	86	18	42
Traverse City, E. end Hannah, Lay & Co's Pier,	44	45	59.74	85	36	53.11
Ontonagon, Light House,	46	52	18.35	89	18	29.46
Houghton,	47	07	15.00	88	33	27.12

The foregoing positions, as far as Vulcan, inclusive, are selected from the numerous determinations of the United States Lake Survey; Ann Arbor has been determined by the Director of the Observatory; New Buffalo and Niles are from Col. Graham's determinations; Monroe, Traverse City, Ontonagon and Houghton are from the Lake Survey Charts, and the co-ordinates of the remaining localities have been calculated from Farmer's large sectional map of the State.

The following table exhibits the difference of time between Detroit and some important points in the State:

TABLE OF LOCAL TIME.

LOCALITIES.	Time slower than Detroit Time.	LOCALITIES.	Time slower than Detroit Time.
	m. sec.		m. sec.
Port Huron,	2 30.57*	Battle Creek,	8 34.99
Pontiac,	0 59.85	Kalamazoo,	10 13.13
Monroe,	1 07.97	Traverse City,	10 18.00
Ypsilanti,	2 18.89	Grand Rapids,	10 20.39
Flint	2 34.35	Allegan,	11 20.94
Ann Arbor (Obs.,)	2 45.35	Grand Haven,	12 47.90
East Saginaw,	3 33.34	Niles (Trinity Ch.,)	12 52.93
Adrian,	3 48.62	Muskegon,	12 53.91
Owosso,	5 03.91	Manistee,	13 05.29
Hillsdale,	5 05.55	Escanaba,	16 00.19
Jackson,	5 22.54	Marquette (L. H.,)	17 18.65
Lansing,	5 53.30	Menominee,	18 12.16
Mackinac,	6 16.01	Houghton,	21 04.18
Marshall,	7 35.05	Ontonagon,	25 04.42
Coldwater,	7 56.64	Mouth Montreal River.	29 59.15

FOREIGN LOCALITIES COMPARED WITH DETROIT.

Greenwich, England,	5 h. 32 m.	9.51	sec.	faster.
N. Y. City, (Custom House,)	36 "	9.31	"	"
Washington, D. C. (Observ.)	24 "	8.51	"	"
Chicago, Ill. (Old Court House,)	18 "	22.34	"	slower.
San Francisco, Cal.,	2 h. 37 "	23.00	"	"

* *Faster* than Detroit Time

The geographical centre of the main land of the Upper Peninsula is on Sec. 35, T. 46, N. R. 25, W., about three miles east of the Peninsula Railroad, in Marquette County. The geographical centre of the main land of the Lower Peninsula is on Sec. 24, T. 13, N. R. 3, W., township, of Coe, Isabella County. The geographical centre of the main land of the entire State is on Sec. 3, T. 21, N. R. 8, W., in Missaukee County. The geographical centre of the entire state, within its political boundaries (including the lake-areas belonging to the State) is on the S. W. ¼ Sec. 30, T. 30, N. R. 11, W, very near Provemont in Lelanau County.*

The extreme length of the main land of the Upper Peninsula is 318.104 miles, and its extreme breadth 164.286 miles. The extreme length of the main land of the Lower Peninsula, from north to south, is 277.009 miles; and its extreme breadth is 259.056 miles. The greatest actual width of the Peninsula, however, measured along a parallel of latitude, is between Forestville, on Lake Huron,

* These determinations have been made by ascertaining the centres of gravity of sheets of paper of uniform thickness, cut to the exact limits of the mapped boundaries of the areas whose centres were sought.

and Little Point Sable, on Lake Michigan. The width here is 197.057 miles.*

The Base Line of the land surveys of the State runs "seven miles north of Detroit" (probably the Old Capitol), and the Michigan Meridian (which rules south to the old territorial boundary) is 84° 37′ west of Greenwich. The land area of the State is 56,457 square miles, or 36,128,640 acres.

There are 179 islands included within the political boundaries of the State, which have an area from one acre upwards. The total area of these islands is 404,730 acres.

The total length of the lake-shore line within the State is 1620 miles. Besides the larger lakes lying upon the frontiers, the State includes within its bounds 5173 smaller lakes, having an area of 712,864 acres. The following Table sets forth the leading data respecting the "Great Lakes."

	Length miles.	Width miles.	Depth feet.	Coast Line in Michig'n miles.	Elevation above sea. feet.	Area sq. miles.
Superior,	460	160	988	524	599.2	32,000
Michigan,	360	108	900	637	580.6	20,000
Huron,	270	160	300	424	580.6	20,000
Erie,	250	80	200	39	565	6,000
Ontario,	180	65	500		262	6,000
	1,520			1,624		84,000

* These dimensions are based on the latitudes and longitudes of the points referred to, and the calculated lengths of the degrees of latitude and longitude in the different positions.

The two natural divisions of the State are distinguished by marked physical characteristics. They are completely cut off from each other by the Straits of Mackinac. The northern is rugged with numerous rocky exposures ; the southern consists of plains, plateaux, gentle undulations and moderate hills, with very few outcrops of rocky strata. The northern peninsula is a mineral region ; the southern, agricultural. The climates of the two peninsulas are as distinct as their locations and their topography ; and, in all statements respecting the climatic features of the State, the two peninsulas ought to be separately treated. The meteorological means for the whole State convey very inadequate impressions respecting either of its natural divisions.

The topographical configuration of the State has been the subject of very careful study. The attempt has been made to collect all the important information obtained in running the various levels for railroad and canal surveys, from 1836 to the present time. The recent progress of these enterprises is so rapid, that it has been impossible to make the tables of elevation absolutely complete, but over 6000 elevations have, nevertheless, been tabulated, which give the height of the surface at every point along the surveyed lines, at which the superficial slope exhibits any considerable change. The planes of reference of the various surveys have been elaborately compared with each other, and all the elevations reduced to the Chicago City Datum—which is low water in Lake Michigan in 1847. These elevations, transcribed upon the map of the State, have served, in the Lower Peninsula, for the construction of a system of contour lines, or lines drawn through points having the same elevation above a given plane. We have undertaken to draw a contour line for every fifty feet of elevation above Lake Michigan, and these are exhibited upon the accompanying map.

Throughout all that portion of the State south of Houghton Lake, this map presents a good general picture of the surface configuration. North of that latitude, the data are insufficient; and the contour lines must be regarded as only rudely approximative. Combining, however, the exact data at hand, with our personal familiarity with the northern portion of the Peninsula, and with the inferrences to be drawn from a good map of the water courses, we have produced results which, with many persons, may be regarded as quite preferable to the absence of all information.

These tortuous lines, to the casual observer, may seem to be very easily laid down, and to possess little interest or value; but every intelligent person will be able to appreciate their importance, and to understand that they represent months of careful labor.

A general glance at the superficial configuration of the Lower Peninsula, reveals a surface swelling gently from the lake shores toward the interior regions. The lake waters are hemmed in by no mountainous barriers or rocky ranges of hills. Generally the lake shores are depressed. This is especially the case around the upper half of Saginaw Bay, and along the region from Lake Huron to Maumee Bay. Yet, in almost all cases, the land rises, within a few miles—sometimes quite rapidly, or even abruptly—to the height of one or two hundred feet above the contiguous lake. Steep, or even precipitous shores, are presented in the northern and eastern part of Huron county; through a large part of Presq' Ile county; around Little Traverse Bay in Emmet county; and throughout Charlevoix, Antrim, Lelanau and Benzie counties; and the statement may be extended to Manistee, Mason and Oceana counties. Sleeping Bear Point in Lelanau county is a bluff of incoherent materials facing the lake, and attaining an elevation of 500 feet. The limestone ridge forming the northern angle of the county, rises somewhat precipitously to altitudes of 200 to 300 feet.* Similar elevations approach the shore of Little Traverse Bay. The Sliding Banks on Hammond's Bay of Lake Huron are 77 feet high, and the rocky cliffs about Point aux Barques, rise to the height of 12 to 20 feet.

Along the border of Lake Michigan, stretches a series of sand-dunes or piles of fine, mostly silicious sand, blown up by the prevailing westerly winds. These attain elevations up to 100 and 200 feet. At Grand Haven, the highest reaches an altitude of 215 feet. This is on the north side of Grand river. The highest on the south side attains an elevation of 205 feet. In the neighborhood of New Buffalo, they reach heights of 30, 40, 50 and 93 feet. Back of these dunes

* For a more particular account of the topography and hydrography of this portion of the State see the writer's *Report on the Grand Traverse Region.* 8 vo. pp. 92 with map, 1866.

the surface is generally depressed, and not unfrequently, occupied by a marsh, a lakelet, a lagoon or an estuary. As a rule, these sands are continually shifting before the wind. They are, accordingly, making constant encroachments upon areas occupied and improved by man. Sometimes, as at Grand Haven and Sleeping Bear, the forest becomes submerged beneath these accumulations, and presents the singular spectacle of withered tree tops projecting a few feet above a waste of sands.

The origin of these sands is in the disintegration or solution of rocks more or less arenaceous, and located along the shores to the windward, or in the bottom of the lake within reach of the agitations of the waters. The liberated silicious grains are either thrown directly upon the beach, or, through a process of bar formation, a new beach rises to the surface, with the characteristic lagoon between it and the original beach. Thus, in many situations, the land is extended lakeward, and, while the sands are encroaching on the landward side, compensation is made by the westward retreat of the sand-laden beach which supplies the encroaching sands.

Proceeding from the littoral belt of the Peninsula toward the interior, we find a region considerably more elevated and better drained than ancient official misrepresentations had led the general public to believe. Though presenting no mountainous districts, and no indications of the agency of forces of upheaval, we have a land-area attaining throughout a large portion of the peninsula, an elevation averaging from 400 to 1000 feet. Erosions, dating back into geological time, have pared down the original surface, and established the existing slopes to the lake shores, and even to the lake bottoms. Later fluviatile erosions have scored deep and broad valleys, which mark off the prominent portions into several distinct regions.

Viewing the Peninsula as a whole, we discover, first of all, a remarkable depression stretching obliquely across from the head of Saginaw Bay, up the valley of the Saginaw and Bad rivers, and down the Maple and Grand rivers, to Lake Michigan. This depression attains, nowhere, an elevation greater than 72 feet above Lake Michigan. This elevation is in the interval of three miles separating the waters flowing in the opposite directions. This spot was chosen, in 1837, as the location for a canal, connecting Saginaw Bay with Lake Michigan. It is obvious, that when the lakes stood at their ancient elevations, their waters communicated freely across this depression, and divided the Peninsula into two portions, of which the northern was an island. This depression, for convenience of reference, may be designated the "Grand-Saginaw Valley."

That lobe of the peninsular swell which lies to the south-east of the dividing belt, has its salient longitudinal axis stretching somewhat arcuately from north-east to south-west through Huron, Sanilac, Lapeer, Oakland, Washtenaw and Hillsdale counties. This, which may be called the South-eastern Water-shed, is not broken through by any of the streams, though it is deeply excavated by the Huron river, in Washtenaw county. Various passes exist across it, and the crest rises in four isolated summits. The Oakland Summit, located in the north part of Oakland county, attains an elevation, on the surveyed lines,* of 529 feet, and gives rise to

* It must be remembered that the following discussion is based on elevations along lines of survey for railroads and canals. Generally, therefore, the numerical values given do not represent the highest elevations nor the lowest depressions.

tributaries of the Flint, Clinton and Belle rivers. The Washtenaw Summit, in the north-eastern portion of Washtenaw county, rises to the height of 394 feet, and over, and gives origin, on opposite sides, to tributaries of the Huron river. The Francisco Summit on the borders of Jackson and Washtenaw counties, is 411 feet high, on the measured lines, and divides the waters flowing into the Huron and Grand rivers. The Hillsdale Summit is located in the centre of Hillsdale county, and attains two culminations, one in the south-western part, between Cambria and Reading, where it reaches an elevation of 613 feet, and another in the north-eastern part, in the township of Somerset. This is, therefore, the highest summit south of the Grand-Saginaw valley. It stretches south-westward into Indiana, and, on the borders of that State, presents a culmination of 546 feet, while, through the southern portion of Branch county, it maintains an elevation of 400 to 500 feet. The Hillsdale Summit stretches also into the south-western part of Jackson county, with an elevation of 450 feet, while a spur 400 feet high, extends to Springport in the north-western corner of the county. From Hillsdale Summit rise the headwaters of the St. Joseph, Kalamazoo and Grand rivers, flowing into Lake Michigan, and the Maumee and Raisin, flowing into Lake Erie. Tributaries of the Maumee and St. Joseph rise within a mile of each other, in the townships of Reading and Allen. Tributaries of the Kalamazoo and St. Joseph rise within half a mile of each other, in the township of Adams; and these two streams approach again within two miles, at Homer, Calhoun county. The head waters of the Raisin are within a mile of those of the Kalamazoo, in the township of Somerset, and those of the Maumee approach equally near in the adjoining township of Wheatland. In the northern part of Somerset are two peaks which, perhaps, constitute the real culminations of the Hillsdale Summit. Here, within an area of two miles by three, we may view the head waters of the St Joseph, Kalamazoo, Grand and Raisin rivers; and an area of four miles square would include, with these, the highest tributaries of the Maumee.

The broad north-westerly slope of the south-eastern Water-shed is intersected by six great rivers—the Shiawassee, the Cedar and Grand, which unite at Lansing, the Thornapple, which unites with the Grand at

Ada, the Kalamazoo and the St. Joseph rivers, all pursuing a general north-westerly course, except the latter, which flows south-westerly to South Bend in Indiana, and thence north-west. The surface between these river valleys rises into a corresponding number of swells. Of these, the one between the Shiawassee and Cedar rivers, lying chiefly in Livingston county, and reaching an elevation of 350 feet, may be regarded as a spur of the Oakland Summit. The Ingham Summit, which is the next, lies between the Cedar and Grand rivers, in the south-eastern portion of Ingham county and the contiguous parts of Jackson, and attains an elevation of 391 feet, on the measured lines. The Grand Ledge Summit, between the Grand and Thornapple rivers, stretching across the northern part of Eaton county and into Ionia, attains an elevation of only 250 feet. The Barry Summit, between the Thornapple and Kalamazoo rivers, is a mass exscinded by Battle Creek from the northwestern prolongation of the Hillsdale Summit. It occupies the south-eastern part of Barry county, reaching, with an altitude of 250 feet, into Eaton, Calhoun and Kalamazoo. The north-western prolongation of this, cut off by Gun river, which unites with the Kalamazoo at Otsego, forms the Kent Summit, occupying the eastern part of Allegan county and the southern part of Kent, and having a culmination of 213 feet in the latter county. Between the Kalamazoo and St. Joseph rivers, is placed the north-easterly-elongated mass of the Cass Summit, which is cut off from the western extension of the Hillsdale Summit by the south-western reach of the St. Joseph river. It covers the north-eastern half of Cass county, extending into the south-western part of Kalamazoo, where it finds a culmination at an elevation of 349 feet, while another culmination in the vicinity of Cassopolis reaches an altitude of 384 feet.

Gathering together the foregoing results, we may here present the following compendious summary:

RELIEF FEATURES IN THE LOWER PENINSULA.

SOUTHERN LOBE.

South-eastern Watershed.	Oakland Summit,		529 ft.
	Washtenaw Summit,		394 "
	Francisco Summit,		411 "
	Hillsdale Summit	Somerset culmination,	600?"
		Cambria "	613 "
		California "	546 "
North-western Slope.	Livingston Summit,		350 "
	Ingham Summit,		391 "
	Grand Ledge Summit,		250 "
	Barry Summit,		250 "
	Kent Summit,		213 "
	Cass Sum't	Oshtemo culmination,	349 "
		Cassopolis "	384 "

That Lobe of the Peninsular swell which lies to the north of the Grand-Saginaw Valley is placed, as a mass, midway between lakes Huron and Michigan, with its northwestern borders crowding somewhat upon the region of Grand and Little Traverse Bays of Lake Michigan. It exemplifies, like the Southern Lobe, a strong tendency to a north-east south-west disposition. The primary division of the Northern Lobe is effected by the valleys of the Manistee and Sable rivers, which take their rise upon the highest summit, and flow thence toward the south-west and south-east into their respective Lakes. The Sable has excavated a valley which, in Wexford county and the western part of Oscoda, sinks from 10 to 100 feet below the highest levels, and in the eastern part of Oscoda, and in Alcona County, 200 to 300 feet below the highest levels—the general plains being 90 to 125 feet above the river The valley of the Manistee (as well as its tributary, the Pine) is similarly sunken in an undulating plateau.

The southern division is bounded south-easterly by a continuous slope toward Saginaw Bay. In its central part it is indented by the hydrographical basin of Houghton and Higgins Lakes. In this rests Houghton Lake at an elevation of 589 feet above Lake Michigan. From this lake, the Muskegon river, the largest of the Peninsula, takes its rise, and, flowing south-westerly, marks the position of a broad deep valley, leaving, on the south-east, an elongated watershed stretching from Mecosta County through Clare and Roscommon, into Ogemaw County. This may be distinguished as the Central Watershed. It presents a general elevation of 700 feet and over, throughout its entire length. The Roscommon Summit of the Watershed, upon the eastern borders of the county by that name, attains an altitude of at least 820 feet, and the Clare Summit, in the central part of Clare County, is believed to attain an elevation of 750 feet.

The prolongation of the Central Watershed toward the north-east is nearly cut off by the South Branch of the Sable river, forming thus what may be designated as the Ogemaw Summit, occupying the region around the junction of the four counties, Ogemaw, Roscommon, Crawford and Oscoda. The culminating point is believed to be about 800 feet above Lake Michigan, while the pass separating it from the Roscommon Summit is not depressed below 625 feet.

To the north-west of the central lakes and the valley of the Muskegon, rise three summits detached from each other by shallow passes. The Crawford Summit, with an elevation of over 700 feet, besides occupying the south-western portion of the county by this name, stretches into Kalkaska, Missaukee and Roscommon Counties. Being bounded on the north-west by the valley of the Manistee, on the north-east by the bifurcated valley of the North and South Forks of the Sable, it is limited on the south-west by interlocking tributaries of the Manistee and Muskegon rivers.

The Wexford Summit, in the south-eastern portion of Wexford county and contiguous portions of Missaukee and Osceola counties, includes Clam Lake, and is believed to attain an elevation exceeding 700 feet. By the Pine river, a tributary of the Manistee, whose higher waters issue from the neighborhood of the head-waters of an affluent of the Muskegon, this Summit is isolated, on the south, from the Osceola Summit, located near the centre of Osceola County, and reaching an elevation of over 700 feet.

The northern division of the Northern Lobe of the Peninsula embraces the most elevated land south of the Straits of Mackinac. It appears to consist of two principal summits separated from each other by the pass which gives place to the head waters of the Thunder Bay river, and one of the affluents of the Sable. The eastern, which we may designate the Oscoda Summit, because located chiefly in the northern part of that county, has an elevation of 800 feet or more. Otsego Summit, to the west, occupies a large part of Otsego county, and is said to attain an elevation of 1200 feet. Within its limits, take their rise the steams which water the Grand and Little Traverse regions, as well as those which find outlet in the vicinity of the Straits.*

RELIEF FEATURES IN THE LOWER PENINSULA.

NORTHERN LOBE.

Southern Division.	Central water-shed.	Roscommon Summit,	820 ?	ft.
		Clare Summit,	750 ?	"
		Ogemaw Summit,	850 ?	"
		Crawford Summit,	700 ?	"
		Wexford Summit,	700?	"
		Osceola Summit,	700?	"
Northern Division.		Oscoda Summit,	800?	"
		Otsego Summit,	1200?	"

* Otsego Lake is represented on the maps as having an outlet into the North Branch of the Sable river. This is an error. There are evidences, however, of an ancient outlet, at a time when the water of the lake stood at a higher level. Recently, moreover, a canal has been dug for "lumbering" purposes, which opens connection into the Sable. The lowering of this lake, like that of Houghton and Higgins Lakes, is one of the numberless evidences of a gradual process of desiccation taking place all over the continent—to the east as well as the west of the Rocky Mountains.

The enumeration of the foregoing 18 Summits, in the whole Peninsula, must not be allowed to produce the impression of any very marked irregularities of surface. The summit-districts are not generally mere hilltops, but level or gently undulating plateaux, through which atmospheric and fluviatile erosions have excavated drainage valleys of moderate depths or with gently bounding slopes. This conformation of the surface exists to a marked extent in the Northern Lobe of the Peninsula. There are, consequently, few precipitous hill-sides, and but very limited regions which cannot be subjected readily to the operations of agriculture.

It may convey, in a more exact form, some idea of the nature of the surface, to present the following statistics of the construction of the Jackson, Lansing and Saginaw railroad. On the 120 miles of the road between Wenona and Otsego Lake, the average amount of earth-work per mile was 10,000 cubic yards; the maximum mile, 40,000, and the minimum, 1,000 cubic yards. The deepest cut is 23 feet; the deepest filling, 28 feet, and the longest cut, 5,000 feet. The total amount of culverting is 500,000 feet, board measure; total amount of bridging, 1,320 feet; number of bridges, 15; longest bridge, 200 feet; highest, 28 feet.

As to the Upper Peninsula, the data accumulated do not enable us to speak with so much detail; and no attempt has been made to lay off contour lines. It seems, nevertheless, appropriate to complete our account of Michigan topography by offering some descriptive statements in reference to that Peninsula.

The region between Lake Superior and the northern bend of Lake Michigan, limited on the east by St. Mary's river, and on the west by the Whitefish river, may be referred to as the Monistique Peninsula, from the large Lake Monistique, occupying nearly a central position in it. The principal portion of the drainage, to the west of this lake, is into Lake Michigan, the water-shed running east and west, by a zig-zag line, within six to ten miles of Lake Superior. East of Lake Monistique, the drainage is chiefly into Lake Superior and other waters east of the meridian of the Straits. The streams, however, throughout the whole interior of the Peninsula, are sluggish, and the regions to the east of Point Iroquois, and about the upper waters of the

Tequamenon, are largely occupied by marshes abounding in peat and bog iron ore.

The southern border of the Monistique Peninsula is lined by ranges of limestone hills, which, in the vicinity of Point Detour, are but slightly elevated, with intervening marshes, but, further west, in the vicinity of Mackinac, attain elevations of 150 to 300 feet. Drummond's Island and the Manitoulin Islands are but the eastward prolongation of the same range of hills, and exhibit elevations quite as considerable as those in the vicinity of the Straits. The cliffs at the eastern extremity of Drummond's Island are over 100 feet high, while the surface toward the interior, rises to the height of 200 and 300 feet. The escarpments of Mackinac Island are 140 feet high, and the central plateau is 300 feet high. Westwardly, the same range of hills extends to Little Bay de Noquet, where, as at Mackinac Island, it presents some strongly marked scenery. In approaching the coast, this elevated limestone region is cut by erosions into innumerable islands ranging in extent from a mere point of rock to several hundreds, or even thousands of acres. These, in the vicinity of Drummond's Island, and Point Detour, become a literal labyrinth with almost inextricable passages.

Toward the north shore, a prominent range of hills begins in the region back of Point Iroquois, and extends in a nearly westerly direction, to the coast of Lake Superior, where it abuts in the famous escarpment known as the "Pictured Rocks," and re-appears in Grand Island with its towering promontories. These smoothly rounded and densely wooded hills attain elevations of 300 to 600 feet above Lake Superior. The streams which break through the range are interrupted by falls. The principal of these is the Tequamenon, which has falls of 40, 45 and 15 feet. The Au Train, eight miles above its mouth, has a fall of 95 feet, and nearer the lake, another of 40 feet.

The immediate shore, between Point Iroquois and the Pictured Rocks, is an alternation of low, sometimes marshy, plains, and rounded sand-hills and promontories. The latter, in the vicinity of Carp river, reach an elevation of 100 feet, while the Grand Sable stands 345 feet above the lake.

The Whitefish river marks the location of a well characterized Valley of Erosion, from one to three miles wide, and bounded by unconsolidated banks 100 to 120 feet above the limestone bottom. The river rises in a series of lakelets within nine miles of the shore of Lake Superior; and the Au Train river, flowing into the latter lake, takes its rise in the same vicinity. Along this valley, the most elevated point is not more than 150 feet above Lake Superior. The writer has elsewhere* suggested that this valley is probably the site of an ancient outlet of Lake Superior, whose waters then passed through Little Bay de Noquet, Green Bay, and the Wisconsin depression occupied by Lakes Winnebago, Horicon and Koshkomona, into the valley of Rock river, and thence to the Mississippi.

Of the region west of Whitefish river, the southeastern portion, between the Menominee river and Green Bay, is mostly a gently undulating surface, presenting a general slope in the direction of the water-courses. This slope, in the south-western part of Delta county, is 430 feet, and near the head waters of the Chocolate river, in Marquette county,

* *American Naturalist*, Vol. IV., p. 505. Through inadvertence, it is stated in the *Naturalist* that the valley is hemmed in by "limestone" cliffs. The cliffs are of unconsolidated materials, though limestone frequently appears in the bottom of the valley. The existence of this valley is not the only evidence that it has been a water-course, since the limestone bed is, in some places, seen to be worn into pot-holes.

550 feet above Lake Michigan. North and north-west of this, is the mountainous district, comprising the Iron and Copper regions, each of which is characterized by its own topography.

The water-shed of the mountainous region strikes in a serpentine course, north-west from the head-waters of the Chocolate river, to within ten miles of the head of Keweenaw Bay, whence it bends, by a course still more serpentine, south-westward to Lac Vieux Desert, on the boundary of Wisconsin. In the first reach of its course, it passes through the midst of the Marquette Iron District. The elevation at Negaunee is 775 feet above Lake Michigan; at Ishpeming, to the west of the water-shed, 865 feet, and at the Champion mine, near Lake Michigami, 1011 feet. The summit, on the Marquette, Houghton and Ontonagon railroad, is 1186 feet. Lake Michigami lies 966 feet above Lake Michigan. The hills north of the lake, reach an elevation of 1215 feet. The greatest elevation on the water-shed is in the vicinity of the sources of the Michigami river, which are 1250 feet above Lake Michigan. The Huron Mountains, east of Keweenaw Point, abut upon the shore of Lake Superior, and rise in rugged eminences which give a marked expression of the mountainous character of the Upper Peninsula. Mount Huron attains an elevation of 932 feet, and other Summits rise from 760 to 887 feet above Lake Superior. The region of the water-shed, south-west from Lake Michigami, becomes first less broken, and then a gently undulating plain, to the Wisconsin boundary.

Keweenaw Point is a rocky ridge, which, beginning with the promontory at the head of the Point, forms a water-shed nearly along the central line. From the base of the Point, the range trends south-west into Ontonagon county. Mount Houghton, near the head of the Point, is 884 feet above Lake Superior, and the range attains nowhere a greater elevation than 900 feet above the lake.

Beyond the Ontonagon river, the Porcupine Mountains may be regarded as a fresh development of the range. Rising somewhat abruptly from the immediate vicinity of the lake shore, they trend at first south-southwest for about 30 miles, whence their course is more westerly. The greatest altitude attained near Lake Superior, is 950 feet; but several knolls further inland, attain elevations from 1100 to 1380 feet above the lake.

In concluding this synoptical sketch of the topographical features of Michigan, it remains to direct attention to one interesting generalization which has not heretofore been pointed out. This is what may be styled the *diagonal system* in the physical features of the State. By this expression it is meant to say that the longitudinal axes of the topographical and hydrographical features of the State, especially of the Lower Peninsula, lie in directions which are diagonals between the cardinal points of the compass. It would extend this paper too far, to point out the leading facts which illustrate and establish this proposition; but it is believed that a brief study of the topographical Chart will render the truth of the proposition apparent. The subject will be elsewhere adequately amplified.

The diagonal system in American physiography is not by any means confined to Michigan. The Maumee river of Ohio, with its tributaries, is a striking reproduction of the Saginaw and its affluents. The Maumee, flowing east-north-east, is fed by the Au-Glaize and St. Mary's, from the south-east, the (little) St. Joseph from the north-east, and the Tiffin from the north-west—the last named, in its higher reaches, flowing from Hillsdale county, Michigan, first south-east and then south-west. In Wisconsin, the north-east south-west basin of Green Bay is prolonged through the Fox river into Lake Winnebago. The same trend is seen in the shore-lines about Chegowawegon Bay, the Apostle Islands and the western extremity of Lake Superior. Even the upper Mississippi, whose general course is meridional, divides itself into a succession of reaches, conforming strangely to the law of diagonism, while, on the other hand, the river and Gulf of St. Lawrence are a further indication that something in the course of events which have fashioned the actual surface, has exerted a greater energy in the direction of the diagonals than in the direction of the cardinal points of the compass.

This is not the place to discuss the causes of this well-marked method in the surface-configuration of the north-west. It would be easy to show that these features sustain relations to the underlying rocky structure. It would be equally easy to demonstrate that they are closely connected with the movements of the continental glacier, which geologists believe to have moved, in the lake region, from north-east to south-west, during the epoch immediately preceding the advent of man upon the earth. But, at the same time, it would appear that these features do not conform exclusively to either set of agencies; and that their actual relation to each may be expressed in the following proposition: *The actual topographical and hydrographical axes of Michigan and the whole lake-region, are the resultant of two forces—a* GLACIAL, *acting from the north-east, and a* STRATIGRAPHICAL, *acting along the lines of strike of the rocky formations.*

As a corollary, we should find that where the rocky formations are most consolidated, the resultant lies nearest the line of the stratigraphical force; and where the rocky formations are little consolidated, the resultant approximates the line of the glacial force.

As a second corollary, physical features determined by causes which have *obliterated* the glacial and stratigraphical trends, do not, necessarily, express relations to either force. Of this kind are the small streams whose courses over the diluvial beds have been determined by post-glacial erosions; and river courses, like the St. Clair and Detroit, marked out across lacustrine or other post-glacial deposits which have concealed the surface-features due to geological structure or glacial erosion.

GEOLOGY.

BY ALEXANDER WINCHELL, LL.D.,
CHANCELLOR OF THE SYRACUSE UNIVERSITY.

A synoptical sketch of Michigan geology will naturally be embraced under three general heads: 1. STRUCTURAL GEOLOGY. 2. HISTORICAL GEOLOGY. 3. ECONOMICAL GEOLOGY.

I. STRUCTURAL GEOLOGY.

The Lower Peninsula occupies the central part of a great synclinal basin, toward which the strata dip from all directions. The basin structure is bounded on all sides by anticlinal swells and ridges. Thus, north of Lake Ontario, Georgian Bay and Lake Huron, is a portion of the great Laurentian ridge whose branches extend from this region toward the north-east and north-west. On the north-west is the elevated granitic and dioritic region stretching from Marquette south-west through northern Wisconsin. On the south-west, south and south-east is a bifurcating gentle swell of the outcropping Devonian and Silurian strata, which stretches southward to Cincinnati and central Kentucky.

The limits of this great geological basin exceed, somewhat, the bounds of the Lower Peninsula, as the centripetal dip can be traced, on the east, as far as London, Ont.; on the west, to Madison, Wis.; on the northwest, to the vicinity of Marquette, and on the north, to the Sault Ste. Marie. Within these limits, the outcropping edges of strata older and older in the series, are passed over in traveling from the centre of the Peninsula outward. The whole series of strata may be likened to a nest of wooden dishes. The great hydrographic features of the region present a striking conformity to the trends of these outcropping strata, as will readily be seen by comparing the longitudinal axes of Lakes Erie, Huron and Michigan, as of Georgian, Little Traverse and Green Bays, with the strikes of the neighboring formations, as delineated on the Geological Map.

The Upper Peninsula is divided by the Marquette-Wisconsin anticlinal into two geological areas. The eastern, as just stated, belongs to the great Michigan basin, while the western belongs to what may be styled the Lacustrine Basin, since Lake Superior covers a large part of its surface. The southern rim of the latter is seen uplifted along Keweenaw Pt., and the south shore of the Lake, while, from this rim, the strata dip north-westerly under the Lake, reappearing in Isle Royale, and, to a limited extent, along the north shore of the Lake. Between the Michigan and Lacustrine basins the metalliferous Marquette-Wisconsin axis interposes a separating belt of about 50 miles.

We here present, in tabular form, a list of the geological formations of the State.

TABLE OF FORMATIONS.

EOZOIC GREAT SYSTEM.		
I. Laurentian System.		
II. Huronian System.		
PALÆOZOIC GREAT SYSTEM, 2680 ft.		
III. Silurian System,	920 ft.	
Lake Superior Sandstone,	300 ft.	
Calciferous Sandrock,	100	
Trenton Group,	60	
Cincinnati Group,	60	
Niagara Group,	250	
Niagara Limestone,		218
Clinton Sub-group,		32
Salina Group,	50	
Lower Helderberg Group,	100	
IV. Devonian System,	1040 ft.	
Corniferous Group,	120	
Little Traverse Group,	200	
Huron Group,	720	
Black Shale,		20
Portage Sub-Group,		500
Chemung Sub-Group,		200

V. CARBONIFEROUS SYSTEM,	720 ft.	
Marshall Group,	160	
Michigan Salt Group,	185	
Carboniferous Limestone,	70	
Coal-bearing Group,	305	
Parma Conglomerate,		100
Coal Measures,		125
Woodville Sandstone,		80
CÆNOZOIC GREAT SYSTEM.		
VI. QUATERNARY SYSTEM.		
Boulder Deposits,		
Modified Drift,		
Lacustrine Clays,		
Bogs, Marls, Dunes, Soils, etc.		

I. The LAURENTIAN SYSTEM.—This System embraces the oldest known stratified deposits. It therefore underlies all the other strata, and its outcropping border constitutes the barrier which limits the great geological basins. Only a small portion of the Laurentian falls within the State of Michigan, and its geographical limits have not been completely determined. It rises in four distinct bosses within the bounds of the Upper Peninsula, three of which lie between the meridians of Keweenaw Bay and Marquette, while the fourth lies south of Ontonagon, on the State boundary, and stretches far into Wisconsin. This System is composed largely of granitic, syenitic and gneissoid rocks. The latter, in Michigan, are known to embrace gneiss, hornblendic and syenitic gneiss, and micaceous and hornblendic schists. Granite proper rarely occurs; syenite is common. It is probable that some of the deposits of white crystalline marble north-west of Menominee will be found to belong to this System.

II. The HURONIAN SYSTEM.—This System comprises the mass of imperfectly stratified rocks with which the great deposits of iron ore, in Michigan, are associated. The area occupied by them stretches in a widening belt from Marquette south and southwestward into Wisconsin,—the Laurentian bosses protruding through the belt. Lithologically, the rocks consist chiefly of diorites, quartzites, chloritic and talcose schists, marbles and iron ores. The results of the most recent studies in these rocks have not as yet been published; but Major T. B. Brooks, under the direction of the State Survey of 1869, '70, determined the usual order of superposition of the rocks, in the Marquette iron region, to be as follows: 1. Laurentian rocks; 2. Quartzite; 3. Talcose schist; 4. Diorite; 5. Ferruginous Quartzite; 6. Diorite; 7. Ferruginous Quartzite; 8. Diorite; 9. Ferruginous Quartzite; 10. Diorite; 11.

Ferruginous Quartzite; 12. Hæmatitic and magnetic ores; 13. Quartzites. The quartzites are whitish or rusty, and vitreous. The diorites are composed of feldspar and hornblende—the feldspar being generally albite, but sometimes orthoclase or oligoclase. These minerals vary greatly in degree of coarseness, but the rock is generally distinctly granular. The talcose schist ranges from an almost pure talc rock to chloritic, argillaceous, and sometimes, micaceous, talcite. The ferruginous quartzites are, in other words, silicious or lean hæmatites, which have been many times mistaken for productive deposits, and so reported, to the great disappointment of investors. The productive iron ores are chiefly hæmatite, undergoing alteration, in contact with water, to red chalk and limonite ("soft hæmatite"), though magnetite occurs abundantly. The mode of occurrence of these ores is not in eruptive outbursts (as formerly supposed), but in vast lenticular masses crowded into the stratification of the schists. Their origin is yet a matter of discussion, but they seem to be masses segregated from the contiguous rocks by some geologic action—perhaps through heat. One of these masses, consequently, is not inexhaustible.

Indeed, several of them have already been worked out.

The Laurentian and Huronian Systems are, in this State, wholly destitute of organic remains, as far as known. For this reason, they were styled, collectively, "Azoic" by Foster and Whitney. It is yet uncertain whether the supposed animal structures found in rocks of Laurentian age in Canada and Massachusetts, are truly of organic origin.

III. The SILURIAN SYSTEM.—1. *Lake Superior Sandstone.*—This is the whitish, reddish, mottled or drab sandstone—sometimes shaly—occurring along the south shore of Lake Superior, from the St. Mary's river to Superior City, and beyond. There has been much discussion in reference to its geological age. The prevailing opinion, at present, makes that portion of it east of Keweenaw Point the equivalent of the Calciferous and Chazy formations of New York. The portion west of the Point is thought by some to be of the same age, while others regard it as the equivalent of the sandstones in Wisconsin and Minnesota, which are generally ranged in the horizon of the Potsdam of New York. As the uplift of Keweenaw

Point has tilted the sandstones on the west, while those on the east have retained their horizontality, there is reason for supposing that the eastern strata are of more recent origin. It may nevertheless be true that the sandstones on both sides of the Point are of the same age, though those on the eastern side were not permanently tilted by the convulsion which upheaved the others. As we find apparently superincumbent strata which answer to the Calciferous, we shall continue to parallelize the Lake Superior Sandstone, presumptively, with the Potsdam.

The geographical extent and surface features of this formation are set forth in the article on Topography. In its geological structure, it consists, as first pointed out by Dr. Houghton, of an upper division composed chiefly of gray sandrock, and a lower division composed of thinner beds of reddish sandrock. These two divisions are further subdivided, as follows:

(f) Sandstone, light colored, thick bedded, rather incoherent, with bands of conglomerate, one of which caps the mass, 20 to 45 ft.

(e) Sandstone, whitish, friable, thin-bedded, with shaly seams, 75 to 100 ft.

(d) Sandstone, white, friable, massive, 50 ft.

(c) Sandstone, laminated, micaceous, banded with white and red layers, alternating with red arenaceous shales, 35 ft.

(b) Sandstone, uniformly red, somewhat argillaceous, thick-bedded, 12 ft.
(a) Sandstone, red or speckled, thick-bedded, hard and coarse-grained, 20 ft.

In the Ontonagon district, the upper division of the sandrock seems to be wanting, and its character is generally laminated or shaly, sometimes becoming decidedly argillaceous.

Keweenaw Point is a ridge of more recent origin than the Huron mountains and the associated upheavals. It consists of an immense dyke or wall of volcanic rock, constituting the axis of the Point, against which rest, on the westerly side, alternating strata of sandstone, conglomerate and bedded dolerite, aggregating many thousand feet in thickness. These features, due to igneous action, continue westward to the Porcupine mountains. Ile Royale marks the site of another igneous outburst, which has also brought into view, along the south-easterly side of the island, a reddish sandstone undoubtedly the equivalent of that upon the southern shore.

No animal remains have been discovered in the Lake Superior Sandstone; but the writer has described some forms of fucoids found near the Montreal river and in the Porcupine mountains, under the names of *Palæophycus arthrophycus* and *Palæophycus informis.*

2. The *Calciferous and Chazy Formation.*—This formation enters the Upper Peninsula at the Grand Rapids of the Menominee river, and, trending nearly north in a broad belt, as far as township 45, bends eastward to follow the strike of the Lake Superior Sandstone, and crosses the St. Mary's river at Sugar Island. Its contact with the underlying sandstone is seen near Munising Furnace, on Sailor Encampment Island and elsewhere; and the limestones of the following group are seen in contact with it at the rapids of the Menominee, on the Escanaba river, on Sailor Encampment Island, and at other localities.

The formation is a more or less calcareous, often coarse-grained, sandstone, with alternations of dolomitic limestone, recalling the prevailing character of the Calciferous formation west of the Mississippi river. The following is a section on the Menominee river.

(f) Limestone, fine, crystalline, evenly and thinly bedded, with argillaceous partings, 4 ft.
(e) Limestone, nodular, concretionary, 3 ft.

(d) Dolomitic limestone, compact, arenaceous, 2 ft.
(c) Limestone, finely argillo-arenaceous, red-banded or variegated with red blotches, in thin and even-bedded layers, 3 ft.
(b) Limestone, hard, dolomitic, with oölitic belts, 4 ft.
(a) Sandstone, white, coarse, with seams of arenaceous shales, 5 ft.

The formation is almost destitute of fossils.

3. The TRENTON GROUP.— These rocks are predominantly calcareous. They form an outcropping belt, about four miles wide, across St. Joseph and Great Sailor Encampment islands, stretching thence westerly in a gradually widening band which bends around to the south-west, lying with its southern border on the west shore of Little Bay de Noquet and Green Bay, and continuing thence across Wisconsin into northern Illinois.

The limestones of this Group form the Little and part of the Grand Rapids, in the Menominee river, and the so-called falls of the Escanaba and other affluents of Little Bay de Noquet, from the west.

The following section from the Escanaba river will serve as a representation of the stratigraphical constitution of the western portion of the Group:

(g) Limestone, impure and dolomitic, in beds 4 to 10 inches thick, 8 ft.

(f) Limestone, thin-bedded, nodular, with irregular seams of argillaceous and cherty matter, 12 ft.
(e) Shales, alternating with arenaceous limestones, highly fossiliferous, 30 ft.
(d) Limestone, light, sub-crystalline, underlaid by dark-blue, crystalline limestone, 7 ft.
(c) Limestone, thin-bedded, uneven, nodular, with silicious veins and concretions, 15 ft.
(b) Dolomitic limestone, thick-bedded and crystalline, 8 ft.
(a) Limestone, greenish-ashen, with concretions, 6 ft.

In the region of the eastern outcrops, we may dispose the strata of the Group into three divisions, as follows:

III. Limestone, light, brittle, breaking with conchoidal fracture, weathering into uneven, wedge-shaped slabs. Highly fossiliferous.
II. Limestone, dark, thin-bedded, nodular, with shaly intercalations. Highly fossiliferous.
I. Areno-calcareous shales, dusky-green or bluish. Abounding in fossils.

Besides the main outcropping belt, an isolated area of horizontally stratified Trenton limestones, 75 feet thick, covering about four square miles, is found about 14 miles northwest of the head of L'Anse Bay. Sulphur island, also, four miles north of Drummond's island, seems to be an uplifted dome of Huronian quartzite, flanked by steeply inclined strata of silico-argillaceous limestone belonging to this group.

4. The *Cincinnati Group* (Formerly "Hud-

son River Group.")—The outcropping belt of these prevalently argillaceous limestones is nearly concentric with the preceding formations, but lies nearer the centre of the geological basin. The strata are well seen on the north side of Drummond's island, in a belt about four miles wide, which extends with equal width across St. Joseph island, and intercepts the southern extremity of Sailor Encampment island. On the northwest, the Group occupies the space between Great and Little Bays de Noquet, forming cliffs 15 to 50 feet high along the shore of the latter. Excavated along its outcropping border to form the basin of Green Bay, it reappears at the southern extremity, and continues in the direction of Winnebago and Horicon lakes in Wisconsin. Dipping from the regions exterior to the Lower Peninsula of Michigan concentrically toward the centre of the Peninsula, the formation underlies the whole of it, making its appearance, on the south, in southern Ohio, and thence to Cincinnati and central Kentucky.

The following section of strata belonging to this Group, is furnished on the east side of Little Bay de Noquet, SE ¼ Sec. 26 T 39 N 22 W:

(f) Limestone, massive, argillaceous, bluish or ashen, and highly fossiliferous, 20 ft.
(e) Blue indurated shale, 2½ ft.
(d) Limestone, very argillaceous and fossiliferous, with irregular patches of shale intermixed, 2½ ft.
(c) Blue shale, greenish on fresh exposures, 5½ ft.
(b) Limestone, very argillaceous, bluish and fossiliferous, 1¼ ft.
(a) Blue shale, greenish on fresh exposures, 6½ ft.

The fine exposure upon the north shore of Drummond's Island presents strata of a similar character, and abounding in beautiful fossil corals.

5. The *Niagara Group.*—This eminently calcareous group of strata forms a belt arching around the northern borders of lakes Michigan and Huron. Constituting the principal mass of Drummond's island, it trends westward, underlying the region west of the southern half of St. Mary's river, and dipping beneath the water of the lakes, where it remains visible, sometimes, to the depth of thirty or forty feet. It is deeply and irregularly eroded along the lake shores, presenting innumerable passes through a labyrinth of small rocky islands. Continuing westward, it underlies the peninsula between Lake Michigan and Big Bay de Noquet, and forming the islands south of Pt. Detour, reappears in the Wisconsin peninsula east of

Green Bay, and follows the coast of Lake Michigan thence to Chicago. Along the shore of Big Bay de Noquet, it rises in picturesque cliffs to the height of 100 and 175 feet.

Eastward from Drummond's Island, the solid masses of this group constitute the Little and Great Manitoulin islands, and reappear at Cape Hurd, to form the peninsula between Lake Huron and Georgian Bay. Thence it strikes south-east to the Niagara river, which gives its name to the group.

In Michigan, the group divides itself into two divisions, consisting of the Niagara limestone above, and the Clinton limestone below. In New York, further divisions are noted. The Niagara limestone, as a whole, may be described as a gray, crystalline, rather fine-grained compact, moderately fossiliferous dolomitic mass, attaining a maximum measured thickness (on Green Bay, Wisconsin) of 217 ft. 10 in. A portion of the mass is generally very thick-bedded, more coarsely crystalline, vesicular, and abounding in *Pentamerus oblongus*, whence it was styled by Dr. Houghton the "Pentamerus limestone." These beds seem generally to occupy a middle position, but observations in the vicinity of Bay de Noquet tend to indicate that the Pentamerus beds are not always in the same horizon. The Clinton limestone is more homogeneous, aluminous and fine-grained, and contains a paucity of fossils.

This group is finely exhibited on the eastern portion of Drummond's island, where the following section of limestone was carefully measured by the writer:

(p) Hard, crystalline, light gray, weathering rough, abounding in *Pentamerus* and corals, constituting the highest ledge,	6.00 ft.
(o) Very thin layers, much broken,	8.00
(n) Rough, crystalline, geodiferous, abounding in *Pentamerus* and corals,	25.93
(m) Concealed slope, which, allowing for dip, makes,	18.87
(l) Gray, crystalline, hård, highly calcareous, burned for lime. Forms upper ledge south of the quarry.	7.00
(k) Areno-calcareous, weathering harsh, abounding in fossils. Uppermost rock seen in the quarry.	5.00
(j) Argillo-calcareous below, resembling (h); areno-calcareous above, resembling (k); weathers unequally; some *Cyathophylloids* at top,	2.75
(i) Dark, coarsely crystalline, exceedingly tough,	.25
(h) White layer, very fine-grained, weathering white, cherty-mottled in the lower part,	3.50
(g) Areno-calcareous with some cherty mottlings; the lower half very hard, the upper, softer and striped with brown,	4.25
(f) Arenaceous, with hard, interlaminated layers; becomes vesicular,	2.00

(e)	Dark gray, very hard, with small geodes. Beautifully ripple-marked at top.	2.00
(d)	Arenaceous, thinly laminated, dark colored; traces of fucoids or branching corals on the upper surface. Gashed with lamellar crystal cavities.	1.25
(c)	Brown limestone, exceedingly tough,	.75
(b)	Dark, areno-calcareous, with alumina disseminated and in wavy streaks,	1.75
(a)	Argillo-calcareous, ashen colored, very fine-grained, thick-bedded—a single stratum being $4\frac{1}{2}$ feet. Contains *Cytherina*, an *Aviculoid* and *Murchisonia*.	6.00
	Distance to the water surface,	1.50
	Total Elevation,	98.30 ft.

North of this locality, lower strata are seen which, added to the above, give us a thickness here of 75 feet for the Niagara limestones, and 32 feet for the Clinton.

On the opposite side of the State, in the vicinity of the Jackson Iron Furnace, on Big Bay de Noquet, the following detail of limestones appears:

(j)	Thin-bedded and argillaceous,	8 ft.
(i)	Talus, sloping back 20 rods,	20
(h)	Coarse, vesicular, massive, fossiliferous,	14
(g)	Very hard, sub-crystalline, fine-grained, compact, flint-like, argillaceous, unfossiliferous, weathering buffish,	8 ft. 2 in.
(f)	Rough, massive, vesicular, fossiliferous,	5 ft. 10 in.
(e)	Fine-grained, crystalline, very compact and hard,	2 ft.
(d)	Rather thin-bedded, banded with argillaceous matter, fine-grained, unfossiliferous,	1 ft. 8 in.

(c)	Hard and sub-crystalline, with conchoidal fracture, unfossiliferous,	11 ft.	
(b)	Rough, vesicular, coarse, weathering into irregular flags or chips, fossiliferous,	5 ft.	
(a)	Fine, hard, sub-crystalline, in beds of 11 to 14 inches, resembling (c),	7 ft.	8 in.
	Total thickness of exposure,	83 ft.	4 in.

The total thickness of the Clinton strata in this part of the State is 38 ft. 10 in., with a persistent, mixed conglomeritic bed of 8 to 12 inches or more, separating them from the Niagara.

The Niagara limestones pass southward beneath the Lower Peninsula, but do not reach the surface within the southern limits of the State, though they have been penetrated in Artesian borings at London in Monroe County.

6. The *Salina Group.* (Formerly "Onondaga Salt Group.")—This is a thin series of argillaceous magnesian limestones and marls, embracing beds and masses of gyspum, and, in some regions, strata of rock salt. It is the lowest stratified rock in the Lower Peninsula. In the Upper Peninsula, its belt of outcrop stretches across the point of land north of the Straits of Mackinac, from Little Point au Chene to near the mouth of Carp river, and following the vicinity of the shore,

from that Point to West Moran Bay. The formation, with the characteristic gypsum, is seen, beneath the water-surface, at the Little St..Martin island, and at Goose island near Mackinac. Dipping beneath the Southern Peninsula, it reappears in Monroe County, where it has been exposed in some of the deepest quarries. Near Sandusky, Ohio, it affords valuable deposits of gypsum. The formation has also been reached in numerous Artesian borings, as at Mt. Clemens, Caseville and Alpena. At the two latter places, a thick bed of rock-salt was penetrated, which is undoubtedly the equivalent of the bed worked at Goderich, on the opposite side of Lake Huron. The total thickness of the formation is not accurately established, but probably, aside from the salt-bed, it does not exceed 50 or 60 feet. The stratification, by combining observations at remote outcrops, may be set down as follows:

III. Calcareous clay, seen at Bois Blanc.
II. Fine ash-colored limestone, with acicular crystals, as at Ida, Otter Creek and Plum Creek quarries, Monroe County, and at Mackinac, Round and Bois Blanc islands.
I. Variegated gypseous marls, with imbedded masses of gypsum, as at Little Point au Chêne and the St. Martin islands.

7. The *Lower Helderberg Group.*— This group of argillaceous and magnesian limestones was not known to the public to exist within the State until announced by the writer in 1870. They form some of the lower portions of Mackinac island and the contiguous shores, and, passing under the Peninsula, outcrop along the western end of Lake Erie, and constitute a large part of some of the islands in that region of the lake. At their northern outcrop, they consist of a series of chocolate-colored, magnesian limestones, more or less argillaceous, occurring in regular layers 4 to 8 inches in thickness, and passing upward by irregular gradations into the brecciated mass of the next group, showing a thickness of perhaps 50 feet. At the southern outcrops, the strata are evenly bedded, rather dark-ashen in color, argillaceous, and lined, sometimes, with darker argillaceous seams. They are often exposed in the quarries of the eastern part of Monroe County, and may be stated to attain the thickness of about 60 feet. They seem to correspond to the Waterlime group of the New York series. The fossils seen in Michigan are *Leperditia alta* and *Spirifer modesta.* *Eurypterus remipes* is also found on Put-in

Bay and other contiguous islands of Lake Erie.

III. The DEVONIAN SYSTEM. 8. The *Corniferous Group.*—This comprises the conspicuous and durable limestone which forms the mass of Mackinac, Round and Bois Blanc islands, and the elevated promontories of that vicinity on both sides of the Straits. It underlies a large part of Emmet and Presqu' Ile counties, and forms the Fox and Beaver islands of Lake Michigan. On the south, it underlies a large part of Monroe County, and stretches southward into Ohio and Indiana. It is the prominent limestone seen at Columbus and Sandusky, Ohio; at Monroe and London, Michigan; and at London and Woodstock, Ontario. This and the Niagara are the great limestone masses which enter into the relief of the physiognomy of the northern states, from New York to the Mississippi river, and furnish sites for the most valuable limestone quarries. The formation everywhere abounds in fossils, and furnishes us, in the form of fish-remains, the relics of the oldest vertebrates which inhabited our planet. Within the limits of Michigan, it is everywhere divisible into two well marked divisions: a lower, brecciated mass, about 150 feet thick, and an upper, somewhat evenly stratified mass, about 100 feet thick. At Mackinac and vicinity the stratification may be generalized, as follows:

IV. Limestone, more or less oölitic, regularly bedded, 25 ft.
III. Limestone, unevenly and thinly bedded, with silicious veins, and cherty nodules, 75 ft.
II. Brecciated limestone, the individual fragments being angular, various in composition, and sometimes little displaced from original juxtaposition, all re-cemented by an indurated calcareous mud, 150 ft.
I. Conglomeritic bed, consisting of a mass of cherty and agatoid pebbles. This occupies the place of the "Oriskany Sandstone" of New York, though not yet identified with it. 3 ft.

In the southern part of the State, the brecciated division presents conspicuous and remarkable features along the shore of Lake Erie, in the vicinity of Pt. aux Peaux and Stony Pt., where it abounds, also, in the mineral strontianite. A generalized section of the group in this part of the State is here presented:

IV. Brown bituminous limestones, seen in most of the quarries of Monroe County; also in Presqu' Ile and Emmet counties, 75 ft.
III. Arenaceous limestone, sometimes resolving itself into beds of friable sandstone and incoherent sand. Monroe County; also Crawford's quarry. 4 ft.

II. Oölitic limestone, as in Bedford and Raisinville, Monroe County, 25 ft.
I. Brecciated limestone, sometimes concretionary, 50 ft.

The Corniferous limestone, as its name implies, abounds everywhere in masses of hornstone. These however, do not occur in all parts of the formation. The very general presence of bituminous matter imparts a prevailing dark color to the rock. This is also frequently seen disposed in very thin partings between the strata. Petroleum often saturates the formation, and, in many places, imparts its characteristic odor. In some localities, it may be seen to ooze from the crevices and float upon the surface of water. Naturally, these manifestations have led to an unlimited amount of confident, but ignorant and wasteful well-boring. In consequence of the more or less shattered condition of the whole formation, streams of water have coursed through it, and worn out extensive subterranean passages and caverns. In these, considerable creeks sometimes wholly disappear; while they serve also, as means of communication between Lake Erie and some of the inland lakes.

9. The *Little Traverse Group.*—This is composed chiefly of the "Hamilton Group"

proper, of the New York geologists; but, as the lower limits of the Hamilton have not yet been clearly fixed upon in this State, we apply the above term to a series of limestones outcropping in the vicinity of Little Traverse and Thunder Bays, and constituting physically a single mass. They have been made the subject of considerable study. In 1860 we made an official survey of the Little Traverse strata; in 1866, a special survey and report; and in 1869 the ground was again officially examined. As the result of all our studies, we submit the following generalized arrangement:

IV. Chert Beds.
III. Buff, vesicular magnesian limestones, overlaid by characteristic crinoidal beds.
II. Bituminous shales and limestones, composed of
  (b) Acervularia Beds above and
  (a) Bryozoa Beds below.
I. Pale-buff, massive limestones, comprising
  (b) Cœnostroma Beds above, and
  (a) Fish Beds below.

The total thickness was set down provisionally at 141 feet, which is probably too low.

This grouping will apparently hold good over extensive regions. The Cœnostroma and Acervularia Beds are extremely conspicuous on the opposite side of the State, while the Acervularia Beds outcrop at Iowa

City, and the Bryozoa Beds at New Buffalo, Iowa. The following is a section of the Cœnostroma Beds near the head of the Bay.

(d)	Dolomitic limestone, pale-buff, very massive, breaking into regular blocks, somewhat arenaceous,	12 ft.
(c)	Dolomitic limestone, similar to above, vesicular, brecciated in places, having a rude concretionary structure,	20 ft.
(b)	Limestone, thin-bedded below, thicker above, broken, with a 10 inch band of dark bituminous soil at top, and thinner ones below,	10 ft.
(a)	Talus, or sloping beach of fragments,	4 ft.

Section of Bryozoa Beds (SE ¼ Sec 1 T 34 N 6 W)

(e)	Limestone, argillaceous, sub-crystalline, the thinner layers shaly, terminated by a few inches of black shale,	14 ft.
(d)	Limestone, very dark chocolate-colored, argillaceous, compact, much broken,	3 ft.
(c)	Limestone, very dark, bituminous, in beds from 6 inches to one foot thick, shaly or subcrystalline,	12 ft.
(b)	Limestone, dark brown, argillaceous, unevenbedded, breaking with a ragged uneven fracture,	5 ft.
(a)	Limestone, dark, compact, argillo-calcareous, breaking with smooth, conchoidal fracture, much shattered,	1 ft.

Section of Acervularia Beds (SW ¼ Sec 2 T 34 N 6 W)

(d)	Shale, bluish, argillaceous, imperfectly seen at top of bank,	2 ft.
(c)	Limestone, varying from dark to light gray, in beds from one to four feet thick, with a rough, somewhat granular fracture. Few fossils.	23 ft.

(b) Limestone, light or yellowish-buff, varying to dark chocolate, argillo-calcareous, breaking with smooth fracture into irregular, sharply angular fragments, rather even-bedded in layers 6 inches to 2 feet thick. In the upper part, alternating with bands of black bituminous calcareous shale and blue clay. The clay beds abounding in beautifully preserved corals. 17 ft.

(a) Limestone, grayish-brown, compact, argillaceous, uneven-bedded, with smooth conchoidal fracture, embracing in its upper part a 4 inch stratum of black, bituminous argillaceous limestone replete with characteristic fossils. 14 ft.

The strata embraced in the above section seem to be the equivalents of the eminently fossiliferous and often argillaceous beds well known at Partridge Point in Thunder Bay, at Widder and Saul's Mills, Ontario, and Eighteen-mile Creek, N. Y., and the less known localities near the head of Cheboygan lake and in Alpena county.*

The belt of strata of the group under consideration, arches across the northern portion of the Lower Peninsula, occasionally outcropping, and everywhere manifesting their proximity by characteristic fossils in the soil (especially *Acervularia*, *Cœnostroma* and *Atrypa reticularis*) and appearing again in

* For further particulars, see "*Report on the Grand Traverse Region*," pp. 40 to 49 and 83 to 97.

extensive exposures in Alpena county, especially along Thunder Bay river, and in the bluffs and islands about Thunder Bay. Without entering here into details of stratification, we may offer the following summarized statements :

The Cœnostroma Beds are extensively developed at the water-surface on Thunder Bay island and contiguous localities.

The Bryozoa Beds are seen immediately overlying, on the island, and at Nine-mile Point, and are extensively exposed along the lower valley of the Thunder Bay river.

The Acervularia Beds are seen in the cliffs along the lake shores north of the mouth of the Bay, and in the interior, in Sunken Lake and at the head of Cheboygan Lake.

The Crinoidal Beds are observed at the top of the bluff on Cheboygan Lake ; and the gray, coarse, magnesian limestones are found in Sunken Lake.

The strata of this age, passing from Thunder Bay under Lake Huron, reappear in Ontario, and passing under the eastern part of Lake Erie, traverse centrally the western half of the State of New York. On Kelley's island, in Lake Erie, generally reputed to be formed of the Corniferous limestone, we find many of the fossils of this group, and it hence appears that the limits of the Hamilton and Corniferous are as obscure here as in northern Michigan.

The Little Traverse group abounds in most interesting fossil remains. Besides a large number of new species, we have signalized the occurrence of three new genera of corals occurring in the northern part of the Peninsula.

10. The *Huron Group.*—(The "Genesee Shale," the "Portage" and the "Chemung" groups of New York.) This series of pre-eminently argillaceous strata constituting a single mass, physically, not only in Michigan, but also in Ohio, Indiana, Kentucky and other regions, we perpetuate by the general designation first employed by us in 1859. These strata underlie extensive areas in the northern and the southern portions of the Peninsula. Northward their outcropping belt strikes arcuately across the Peninsula between the regions south of Grand Traverse and Thunder Bays; while, southward, a considerable part of Allegan, Van Buren, Kalamazoo, Branch and Lenawee counties is underlaid by it. The physiognomy of these

regions is generally plain without rocky outcrops.

The general features of the group may be stated as follows:

The *Black Shale*, at the bottom, attains a thickness of perhaps 20 feet. It is sometimes laminated and fissile, but frequently somewhat massive and indurated. It is a very persistent formation, known, with increasing thickness, in Ontario and all the western States east of the Mississippi river. In Michigan, we see it outcropping in Grand Traverse Bay, on Pine Lake, on Sulphur island of Thunder Bay, on the coast east of Pt. aux Barques, at several localities in Sanilac and St. Clair counties, and in Kalamazoo and Branch counties; and it is pierced in numerous Artesian borings. It is often mistaken for a coal shale, or even for coal itself; but, though it blazes in a fire, its geological position is far below any valuable coal deposits.

The *Portage shales* come next in order above, and consist of a series of whitish and greenish, more or less calcareous shales and clays almost wholly destitute of fossils, and attaining a thickness of probably 500 feet. They outcrop at several points around the shores of Grand Traverse Bay, and again, extensively, at Port Hope and other localities on Lake Huron, southward. They are frequently encountered in river bluffs and artificial excavations in the southern part of the State. Nodules of Kidney Iron ore are everywhere characteristic of the formation, as may be seen at Coldwater and Union City. At the latter place they have been worked for iron. These nodules are found abundantly in the surface deposits of the whole southern portion of the Peninsula. At Pt. aux Barques, the shales are seen to be interstratified with thin beds of crystalline and fossiliferous limestone; and such strata are also encountered to a limited extent in the Artesian borings further south.

The *Chemung Shales*, following next in ascending order, cannot be sharply distinguished from the preceding. They may be assigned a thickness of 200 feet. If these are more, the Portage shales are correspondingly less, since the thickness of the two has generally been found in Artesian borings, to attain about 700 feet. Toward their upper portion, they become more arenaceous, and terminate in a series of laminated, argillaceous, micaceous, friable sandstones, and pass into the lower beds of the next group.

IV. The CARBONIFEROUS SYSTEM. 11. The *Marshall Group.* This arenaceous and generally ferruginous series of strata corresponds to the upper or fossiliferous portion of the "Waverly Sandstone Group" of Ohio, and is probably represented by the Catskill Group of New York, as now restricted. It answers, probably, to the upper portion of the Old Red Sandstone series of Scotland, which, with the Catskill, seems to occupy the base of the Carboniferous System. The Marshall Group was, for a long time, confounded with the Portage and Chemung, but was assigned a distinct place and designation by the writer, in 1859 and 1860. It is seen outcropping in the sandstone bluffs of Pt. aux Barques; and, trending thence southward through Huron, Sanilac, Oakland and Washtenaw counties, it forms the southeastern watershed of the Peninsula. In the southern part of Jackson and most parts of Hillsdale county, it rises in frequent outcrops, and is not unfrequently worked as a quarry stone. It is mostly a somewhat friable rock, with a reddish, buffish or olive color, though in some regions becoming gray or bluish gray. The coloring ferruginous matter is very often arranged in imperfect concentric layers, presenting, on a large scale, a rude concretionary structure. At Battle Creek, it becomes decidedly calcareous, and thence toward the northwest, this consolidating constituent causes the formation to present a marked contrast with the friable condition of the rock on the eastern side of the Peninsula.

This formation is generally rich in fossils, though they exist, chiefly, in the form of casts and impressions. Marshall, Battle Creek, Holland, Pt. aux Barques and numerous localities in Hillsdale county, are quite productive. Fish and crustacean remains are not abundant. The molluscous fauna embraces many species of *Nautilus*, *Goniatites*, *Orthoceras*, *Bellerophon*, besides *Nuculana*, *Solen*, *Cardiomorpha*, and many other genera. Brachiopoda, except *Rhynchonella*, are of infrequent occurrence.

12. The *Michigan Salt Group*.—This is eminently argillaceous, but the included stores of Gypsum and brine confer upon it a great degree of commercial importance. Stratigraphically, it consists of beds of clay and shale, with thin intercalated strata of limestone, and an apparently persistent bed of gypsum, having a thickness of ten to twenty feet. As the group embraces no porous stra-

tum capable of serving as a reservoir of the brine, no considerable supplies of brine are obtained in the formation, but they occur in the underlying sandstones of the Marshall group.

This Group outcrops characteristically, near Grand Rapids, and, on the eastern side of the State, on the shore of Saginaw Bay, at Alabaster. At both these localities the gypsum is extensively worked. Evidences of the persistence of the gypseous deposits are known to exist many miles toward the interior. The formation becomes excessively thinned in the southern bend of its circuit.

The only fossils discovered in the Group are obtained from Alabaster. They present marked affinities with the fauna of the Carboniferous limestone; and the writer entertains little doubt that this Group is a mere local condition of the lower portion of the Carboniferous limestone. This phenomenon is understood to be reproduced in Nova Scotia. Thickness about 185 feet.

13 The *Carboniferous Limestone.*—This formation answers to some portion of the great calcareous deposits of the Mississippi Valley, which, for that reason, might be styled the Mississippi River Group. In Michigan, it

outcrops quite frequently in Spring Arbor and neighboring portions of Jackson county, and very extensively at Bellevue and Grand Rapids, and, on the opposite side of the Peninsula, at Pt. au Gres, the Charity islands and Wild Fowl Bay. In the eastern outcrops, it presents a mass of calciferous sandstone at bottom, while elswhere the formation is almost exclusively calcareous. At Grand Rapids it encloses a stratum of red ferruginous, argillaceous limestone, five feet thick, which, like some of the other argillaceous strata, possesses hydraulic properties. Total thickness does not exceed 70 feet.

The limestones are generally quite fossiliferous. *Lithostrotion Canadense* may be regarded as indicating some representation of the St. Louis member of the Mississippi Valley limestones, while *Spirifera Keokuk*, and perhaps other forms, establish the existence of the Keokuk member. It is probable that the highest member—the Chester limestone, is unrepresented, as in other northern regions, while the lowest member may yet be shown to be present in the Michigan Salt Group.

14. The *Coal-bearing Group.*—This occupies the central portion of the Peninsula, ex-

tending from Jackson on the south, to Town 20 on the north, and from Range X. west, to Range VIII. east, of the meridian. We may distinguish three members, as follows :

(a) The *Parma Conglomerate* is the well-known "Conglomerate" of the coal regions of Ohio and other western States. This is the oldest geographical designation bestowed upon the formation, and is derived from Parma, Jackson county, where it outcrops in a quarry of whitish, glistening, somewhat friable, massive sandstone, with scattered pebbles. It attains a somewhat uniform thickness of 100 feet.

(b) The *Coal Measures*, consisting essentially of a series of carbonaceous shales, sandstones, clays, and one persistent bed of bituminous coal, from three to four feet thick. To these are added local beds of black-band iron ore and considerable Kidney ore, though neither ore possesses economical importance in Michigan. The total thickness of these measures does not exceed 125 feet. The following is an average section :

V. Bituminous shales and clays, 40 ft.
IV. Black-band, passing into black limestone, 2 ft.
III. Bituminous and cannel coal, in one or more seams, with aggregate thickness of 3 to 11 feet.
II. Fire-clay and sandstone, 23 ft.
I. Shale, clay, sandstones and thin seams of coal, 50 ft.

(c) The *Woodville Sandstone*, a persistent deposit, presenting variable characters, but generally more or less friable, ferruginous and gritty. At Woodville it is buffish in color, in Shiawassee county, buffish-gray, in Ionia county, red and gray-mottled. Thickness 80 feet.

The Coal-bearing Group of strata presents no general dip. Their normal position is nearly horizontal; but local dips are of frequent occurrence. Slight geological disturbances have caused numerous anticlinal ridges on which the denudation which leveled the country has worn to a greater or less depth—sometimes leaving the Woodville sandstone at the surface, sometimes exposing the coal measures, and at other times even bringing to view the Parma Sandstone. Hence the local details of the geology within the geographical bounds of the Group, are exceedingly complex and difficult to settle.

V. The Quaternary System.—The surfaces of the Laurentian, Huronian and Palæozoic rocks above described are overlaid generally by a sheet of unconsolidated materials consisting of clay, boulders and sand, with frequent superincumbent beds of marl and peat. Along the shores of the great lakes, these are seen to consist chiefly of strictly and horizontally stratified clays, mostly of bluish and coppery colors. In the interior, we find partially and obliquely stratified, alternating beds of sands and clays, with occasional courses of boulders. In the Northern Peninsula, and, to a great extent, in the Southern, a bed of wholly unstratified clay and rounded boulders rests immediately upon the rocky surface. The superficial beds of marl and peat, with not infrequent bogs of iron and manganese, connect the history of the past with the present. The sand dunes along the lake shores are merely piles of sand blown up by the winds, as explained in the article on topography.

2. HISTORICAL GEOLOGY.

The first land within the limits of the State, was that which we have mapped as Laurentian and Huronian. No part of the existing continent is older; while nearly all other portions were still sea-bottom, except an angulated belt north of the Great Lakes and the river St. Lawrence. This original area has been subjected to a vast amount of subsequent erosion, and correspondingly diminished in its elevation and contracted in its dimensions. Its upheaval marked the close of Eozoic, and the dawn of Palæozoic, Time.

At the end of the first period of Palæozoic Time, an igneous outburst called into existence Keweenaw Point, the Porcupine Mountains; and the intervening copper ranges, together with Ile Royale and limited areas, upon the immediate shore of Lake Superior.

After this time, there were no local disturbances of special importance. The whole continental mass, east of the Rocky Mountain region was, by degrees, bodily uplifted. The Michigan region slowly emerged. The valley which was to become the basin of Lake Superior, was, at first, a bay of salt water. With the progress of continental upheaval, it became isolated from the sea, and was, for ages, a salt lake. The sea still set up the valley of the St. Lawrence to the head of the present hydrographical basin of Lake Ontario.

At the end of the Silurian Age, the whole Upper Peninsula had emerged, but the Lower Peninsula was still sea-bottom. On the west, the continent reached down to Chicago, and, on the opposite side, its shore trended southeast to London and the Niagara river. At the close of Devonian Time, the Lower

Peninsula marked the position of a vast bay opening southward. It is not certain whether the anticlinals on the south of the Peninsula had an existence at this early period, or not. It is more probable that the coal-making marshes of Michigan were continuous with those of Ohio, Indiana and Illinois, but this is far from certain.

At the end of the Carboniferous Age, all Michigan was dry land. But none of the great lakes existed, except Superior. The region which is now the centre of the Lower Peninsula was probably less elevated than the regions which now lie upon the borders and in the beds of Lakes Huron and Michigan. The surface denudations going forward through Mesozoic and Cænozoic Time, isolated the coal regions of Michigan and Ohio, if they were ever connected, depressed the regions which were to become the basins of Lakes Michigan, Huron and Erie, and excavated the first Niagara gorge. The drainage of the great northern sea changed it to a lake of fresh water, in which rose the St. Lawrence, flowing into the Atlantic, and probably another great stream flowing through the hydrographical basin now occupied by Lake Michigan and the Illinois river, to the Mississippi and the Gulf of Mexico. No traces of the Flora and Fauna of Michigan, during this long period, have been preserved; but without doubt, forms of animal and vegetable life adapted to the physical situation, were abundant.

At length the region which was to become Michigan, was buried, in common with the entire northern part of the continent, beneath a burden of accumulated snow and ice. This, like modern glaciers, underwent a slow motion which imparted a grinding action to the sheet of ice, and materially modified the surface features of the underlying country. The direction of this movement, on the eastern side of the State seems to have been from the north-east; on the western side, it may have been more from the north. The erosion of the continental glacier gave origin to the boulders and finer materials which occupy the present surface, and its movement transported them southward. By such action were deepened, if not originated, the valleys of Lake Erie, Lake Huron, Saginaw Bay and Lake Michigan with its appended bays. This action, combined with the strike of the underlying strata, has determined those trends in the physiographic features of the State, which we have designated as the "Diagonal System."

In due time, a change of climate, dissolving the glacier, originated torrents of water which imparted an imperfect stratification to the superficial portion of the drift materials. There was, perhaps, a subsidence which buried the whole State again beneath the waters of the ocean. Whether this were so or not, the great valleys excavated by Mesozoic and glacier agencies, were left filled with the water which either was, originally, or in time became, fresh water. The breadth of the great lakes exceeded vastly their present dimensions. Lakes Erie, St. Clair and Huron were one. Through Saginaw Bay and the valley of the Grand River, Lake Huron connected with Lake Michigan. The latter spread over the prairie region of Illinois. By the removal of the eastern barriers, the lakes were slowly drained to their present dimensions.

The surface of the Lower Peninsula was, at first, dotted with almost numberless small lakes. Many of these, by filling with sediments, marl and peat, have become converted into marshes or even meadows and arable lands; and the remainder of them are undergoing the same process.

It is likely that in America, as in Europe, man made his appearance while the dissolution of the continental glacier was in progress. We have, at least, some evidence of his presence in Illinois, while the prairies were a lake-bottom.

3. ECONOMICAL GEOLOGY.

As the commercial statistics of Michigan are presented in a separate article, we shall content ourselves, in this connection, with little more than a catalogue of the economical products of the geology of the State.

I. METALS AND THEIR ORES. 1. IRON.—(a) *Hæmatite* and *Magnetite*, in immense lenticular masses of unsurpassed purity, in the Huronian rocks of the Upper Peninsula. The Hæmatite presents itself as granular, slaty, micaceous, specular, crystalline, and earthy. Under the action of water, it becomes soft hæmatite and red chalk, and by a chemical union with water, assumes the character of Limonite, which is also styled by the miners, soft hæmatite. It also occurs to a limited extent in crystalline forms. The magnetite is generally massive and granular, with distinct crystallizations, which are sometimes also disseminated through the contiguous chloritic schists. (b) *Limonite*, altered from the Huronian hæmatites, as an earthy ore or ochre, or, not unfrequently, re-deposited in stalactitic, mammillary, botryoidal and velvety forms of great beauty. Limonite occurs, also, in immense quantities, and widely distributed over the State, in the forms of bog ore, shot ore, yellow ochre, or even in some cases, massive rock-like beds. (c) *Kidney Ore* abounds in the Huron clays, presenting, like the bog ores, various degrees of purity, and, like them, employed to a limited extent, for iron-making. (d) *Black-band* in the Coal Measures, but not known to possess economical importance.

2. COPPER.—(a) *Native* in the "trap" of Lake Superior, in sheets, and strings, and masses; also in certain conglomerates and grits associated with the beds of trap, where it occurs in grains and in powder, like the other detrital materials. This is its condition in the famous so-called "Calumet Vein," also in parts of the Porcupine mountains. (b) *Chalcopyrite* or Copper Pyrites and other ores, in the Eozoic and other metamorphic rocks. While these ores sustain an important industry in the dominion of Canada

(Bruce and Wellington mines) native copper is the chief resource in Michigan.

3. SILVER.—(a) *Native*, existing, to some extent, in most of the native copper, and not unfrequently associated with it in a state of purity. (b) Existing as a *vein ore*, in limited abundance, in the trappean rocks; and, at Silver islet (Canada) and vicinity, developing an important special industry. Also, as a sulphide in union with galena, in the dolerites of Lake Superior, but not existing to any important extent.

4. LEAD.—*Galena* in unimportant and unpromising veins in the dolerites.

5. GOLD.—*Native*, existing, to a limited extent, in the Lake Superior region.

6. MANGANESE.—(a) In connection with certain hæmatites of Lake Superior. (b) In numerous bogs in the Lower Peninsula, where it is sometimes used as a black pigment.

II. SALT.—Occurring in the form of brine which has its origin in three different formations: 1. The *Salina Group*, which underlies the Lower Peninsula, and has been pierced and found to afford brine at Port Austin, Caseville, Mt. Clemens, Jackson, Lansing, Grand Haven, Alpena and other localities.

Only in the first three does the supply sustain the manufacture of salt. At Alpena and Caseville rock salt occurs as at Goderich. 2. The *Michigan Salt Group*, which supplies most of the wells along the Saginaw river and vicinity, and affords a brine of remarkable strength, but containing considerable chloride of calcium which, nevertheless, as manipulated, does not interfere seriously with the manufacture of salt. These wells average about 800 feet in depth, and pass through the whole thickness of the coal-bearing group to the Marshall sandstone, into which the brine descends and accumulates. The brine is obtained from these Artesian borings by pumping. 3. The *Coal Measures.* Some of the shallow wells in the lower portion of the Saginaw Valley are supplied from this source with weaker but purer brine than that obtained from the group below. The Parma conglomerate serves as the reservoir for this group of salt-bearing strata. It may be added that the dish-like conformation of the strata of the Lower Peninsula, preventing the passage of water from side to side, retains the soluble constituents of the rocks, and hence they are all somewhat saliferous.

III. MINERALS USED IN CERTAIN

CHEMICAL MANUFACTURES.—1. The BITTERNS rejected in the salt manufacture, are now extensively employed in the production of soda. 2. IRON PYRITES occurs in the Huron Group in such abundance as to promise availability, at some future time, in the process of alum-making. 3. LIMESTONE suitable for fluxing, occurs in unlimited quantities in the Trenton and Huronian rocks of the Upper Peninsula, as also in the form of calc-spar veins in the cupriferous region. In the Lower Peninsula the limestones of the Little Traverse and Corniferous Groups are equally available.

IV. MINERALS USED IN AGRICULTURE.—1. GYPSUM in remarkable abundance, purity and beauty, in the *Michigan Salt Group*, at Grand Rapids and Alabaster. Occurs also in the *Salina Group* at Little Pt. au Chene, and may be found, perhaps, in Monroe county. 2. MARL, generally distributed, and occuring at the bottom of lakelets and marshes. 3. PEAT, as the uppermost layer on the sites of filled lakelets, and around the low borders of existing lakelets.

V. MINERALS USED AS PIGMENTS.—1. IRON and MANGANESE OCHRES, in bogs and marshes through the Lower Peninsula and the Monistique Peninsula. 2. Ferruginous shales.

VI. COMBUSTIBLE AND CARBONACEOUS MATERIALS. 1. Coal underlying about 6000 square miles of the central portion of the Lower Peninsula. Generally bituminous and of the character of the average Illinois coals. Cannel coal exists to some extent, but has not yet been developed. The principal coal mines are at Corunna and Jackson. At Grand Ledge and other points, the facilities for mining are equally good. The undisturbed condition of the strata has left the coal deposit generally so low that drainage of the mines is impracticable except by pumps. 2. Bituminous Shale, in the Huron Group, capable of furnishing oil, gas, stearine &c. 3. Petroleum in the Huron Shales; but which, from the absence of anticlinal axes and overlying porous strata, has not accumulated in reservoirs. 4. Peat, in bogs, throughout the State.

VII. REFRACTORY MATERIALS. 1. Sandstone. 2. Fire-clay of superior quality, in the Coal Measures. 3. Moulding sand: (a) *White*, in the Corniferous Limestone of Monroe County; (b) *Colored*, in the drift.

VIII. MATERIALS FOR BRICKS etc. 1. CLAY, in the Huron Group (as at Coldwater) and in the lacustrine deposits and the ordinary drift, suitable for (a) *Common Bricks* and pottery, (b) *Buffish* (or "Milwaukie") *bricks*, and even *white* bricks and pottery, as at Spring Lake. 2. WHITE SAND of superior quality for glass, in Monroe county, and in the Woodville Sandstone of Jackson county.

IX. MATERIALS FOR CEMENTS AND MORTARS.—1. HYDRAULIC LIMESTONES, in the Salina and Lower Helderberg Groups of Monroe County, and probably, also, in the Hamilton of Alpena County and elsewhere; also, in the Michigan Salt Group of Grand Rapids and Alabaster. 2. STONE FOR QUICK-LIME, in great abundance. Used extensively from the Corniferous, in Monroe County, and from the Carboniferous, in Eaton and Kent counties. 3. PLASTER, in the Michigan Salt Group and the Salina Group.

X. GRINDING AND POLISHING MATERIALS.—1. GRITSTONES, of superior quality, from the Marshall Group at Pt. aux Barques, and coarser ones at Napoleon. The Huron grindstones have a national celebrity. 2. HONESTONES, in the Huronian strata near

Marquette, from the silicious schists. 3. POLISHING POWDERS, in the drift in many places.

XI. BUILDING MATERIALS.—1. GRANITE, SYENITE, DIORITE, GNEISS, etc., equal to any in the world, in the Upper Peninsula. 2. ROOFING SLATES, in the vicinity of L'Anse and at other points. 3. SANDSTONES: (a) *Brown freestone*, somewhat reddish or mottled; otherwise very similar to the Portland (Ct.) brown sandstone. Occurs near Marquette, and somewhat inferior qualities at many other points in the Upper Peninsula. (b) *Reddish* and *mottled freestone*, from the Woodville formation at Ionia and vicinity. (c) *Bluish* and *gray freestone*, at Pt. aux Barques—same as Cleveland stone. (d) *Buffish freestone*, at Napoleon and Hanover, Jackson County. (e) *Whitish freestone*, in Parma formation at Parma. 4. LIMESTONES, in the Corniferous, at London, Monroe County, and in Presqu' Ile County; also in the Hamilton in Little Traverse Bay; also in the Niagara of Drummond's island and Little Bay de Noquet—the same as at Lockport, N. Y., and Joliet, Ill. 5. SAND and GRAVEL, from the drift. 6. BOULDERS, from the drift, extensively used for foundations, and even sometimes for superstructures.

XII. MATERIALS FOR ORNAMENTAL PURPOSES.—1. MARBLES: (a) *Statuary* in the Menominee region. (b) *Mottled* and silicious in the Huronian of Marquette County. (c) *Coralline* from the Little Traverse Group of Presqu' Ile and Alpena counties. 2. ALABASTER, variously colored, from the Michigan Salt Group of Grand Rapids; also, *white* and *clouded*, from the same group at Alabaster. 3. PRECIOUS STONES. *Agates:* banded, fortification and moss agates; *jasper*, *chalcedony*, *chrysocolla*, *chlorastrolites*, etc.,—all in the doleritic rocks of the Upper Peninsula.

XIII. MINERAL WATERS. 1. SALINE WATERS.—(a) *Brines*, used for salt-making, as before stated. (b) *Medicinal*, of insufficient strength for salt-making, but containing carbonate and sulphate of potash, soda and iron, with sometimes traces of lithia and other ingredients, occurring in the form of springs, as at Ann Arbor, St. Joseph and other localities, or obtained by boring, as at St. Louis, Lansing, Spring Lake and many other points. 2. CARBONATED waters, with more or less of soluble salts, as at Eaton Rapids. 3. SULPHUR WATERS, issuing in springs, as occurs most copiously at

Raisinville and on the shore of Lake Erie in the town of Erie, Monroe County; also at Ann Arbor and many other points. Also, issuing from Artesian borings, especially in the Corniferous limestone and the Huron Group. As before remarked, the conformation of the strata has retained all their original soluble constituents; hence, all Artesian waters in the State, save some outlying, leached-out patches of the Parma sandstone, will be found mineralized. The so-called "Magnetic" waters of the State are not themselves magnetic; but marked magnetic phenomena manifest themselves about the wells. These certainly arise, in part, through induction from the earth, without regard to the waters; but some experiments seem to indicate a power of *excitation* of magnetism possessed by the waters themselves.

XIV. MISCELLANEOUS. 1. LITHOGRAPHIC STONES, of coarse quality, in the Clinton and Salina Groups. 2. STATIONERS' SAND, Magnetic iron-sand assorted by the waves upon the lake-beaches. 3. PAVING STONES from the drift.

CLIMATE.

BY ALEXANDER WINCHELL, LL.D.,
CHANCELLOR OF THE SYRACUSE UNIVERSITY.

THE meteorology of the region of the "Great Lakes" is singularly interesting, and is, also, closely connected with the industrial resources and the civilization of that portion of our country. We have, accordingly, bestowed upon this subject, a large amount of study, some of the general results of which will be embodied in the present paper. Our investigations have extended to all the elements of climate—temperature, pressure, moisture, precipitation, cloudiness, winds and occasional phenomena; and we have compiled voluminous tables giving mean monthly results for series of years at a large number of localities, both within and without the State of Michigan. Our tables and results represent all the meteorological observations ever published from within the limits of the State, as well as many observations yet unpublished. For purposes of comparison, we have collected similar data, respecting more than fifty selected localities lying outside of the State of Michigan. The Michigan observations aggregate 284 years, and those of other localities, 493 years.

In the present paper we direct especial attention to the subject of temperature; and, instead of offering a body of statistical tables, we present the reader a series of *isothermal charts*, which, with the explanatory remarks with which we accompany them, will exhibit intelligibly to the eye, the general thermometric features of the different parts of the State. For the purpose of exhibiting a comparison between the climate of Michigan and that of the states contiguous, on the west, we have extended the territory covered by these charts, as far west as the Missouri river, and as far south as Springfield, Illinois. The sinuosities of the several lines will demonstrate, at a glance, the peculiar character of the climate of Michigan, and the fact that, both in summer and winter, it is better adapted to the interests of agriculture and horticulture, and probably, also, to the comfort and health of its citizens, than the climate of any other northwestern state.

The marked peculiarity of the climate of Michigan, in these respects, is attributable to the influence of the Great Lakes, by which the State is nearly surrounded. It has long been known that considerable bodies of water exert a local influence in modifying climate, and especially, in averting frosts; but it has never before been suspected that Lake Michigan, for instance, impresses upon the climatic character of a broad region, an influence which is truly comparable with that exerted by the great oceans. That such is the fact will become apparent when we turn our attention, for a few moments, to the charts.*

We take first into consideration the chart or set of curves for July. Each of these curves—that for 73°, for instance—passes through all the places having the same mean temperature for the month of July. The mean July temperature for several places

* We think it will be conceded that the present writer was foremost in bringing into notice these great climatic facts. The conclusions of this paper were first foreshadowed in a *Report on the Grand Traverse Region* in 1866, and a paper read, the same year, at the Buffalo Meeting of the American Association for the Advancement of Science, entitled "*The Fruit Belt of Michigan.*" The subject was followed up in a carefully elaborated memoir on *The Isothermals of the Lake Region* read at the Troy Meeting of the Association, in 1870. This paper was appended to the writer's *Report on the Progress of the State Geological Survey*, 1870; and an abstract was published in the *Journal of the Austrian Society for Meteorology*, at Vienna, Vol. VII. p. 351, *et seq.*

along each curve, has been determined from good observations continued through a series of years; and the July means for the places between the principal ones, along the curve, are reasonably assumed to be the same as those of the principal places.

Turning our attention, then, to the curves, or isothermal lines for July, we are at once impressed by the magnitude of the deflections of the isothermals in passing the great lakes. These deflections are toward the south, in consequence of the cooling influence of the lakes. In the presence of that influence one must pass to a more southern latitude to find the same degree of warmth as exists in the regions removed more or less from the influence. In the lower peninsula of Michigan, the lines all form loops opening southward, showing that the mean temperature of July, in the interior, is much higher than along the lake borders. And yet, within the peninsula of Michigan, the isothermals do not attain so high a northern limit as in the continental region west of Lake Superior. The isotherm of 70°, for instance, first appears within the limits of the chart in the latitude of 48°, in the valley of the Red River of the North. Passing southeasterward and eastward to the valley of the Menominee river, it comes within the influence of Lake Michigan, and bends directly southward through Green Bay and Milwaukie to latitude 42° 40′, and thence trends northward to Traverse City, in latitude 44° 40′. Here it is deflected southward again, under the influence of Lake Huron, and, passing Saginaw and Sanilac, finally bends north-eastward to attain its normal position, striking Penetanguishene on Georgian Bay of Lake Huron. West of Lake Michigan, this isotherm sweeps across a latitudinal belt of five and a half degrees. Within the peninsula of Michigan, it is deflected first northward two degrees, and then southward one and a half degrees.

Similar deflections are experienced by the isotherms between 67° and 72°. The isotherms of 73°, 74°, and 75° appear to escape much of the influence of Lake Huron. The isotherm of 74° divides in southern Michigan —one branch passing eastward through northern Ohio and the other through central Indiana and southern Ohio. The state of Ohio, consequently, constitues an area of uniform temperature in July, which is identical with the mean temperature of central Michigan to the limit of four and a half degrees of latitude, or 300 miles, further north.

An area in the southeastern part of the peninsula of Michigan seems to be an area of cold; since the temperature is two or three degrees colder than it is on either side. There exists a region in this part of the State which is topographically elevated about 300 feet above the general level of the peninsula. It is the region of outcrop of the sandstones of the Marshall Group, but it is not entirely coincident with this area of cold. An area of warmth seems to be indicated in northern Iowa.

It will be observed that the cooling effect of Lake Michigan is somewhat greater on the west side than on the east. Not only are the isotherms deflected from a higher latitude on the west side, but they likewise attain a somewhat lower latitude. The lowest deflection of the curve of 75°, for instance, is at Ottawa, Ill., to the west of the meridian of the lake. The curves of 71° and 72° are also somewhat more southern on the west side than on the east. This circumstance is undoubtedly accounted for by the slight preponderance, during July, of winds from the east of the meridian. Thus, at Chicago, this preponderance is as 60 : 33=1.82; at Milwaukie, as 47 : 38=1.30. But at Milwaukie and further north, northerly and even northwesterly winds feel the influence of Green Bay.

Contrasting with these results those represented on the isothermal chart for January, we are at once struck with these phenomena: 1st, the great deflection of the isothermal lines; 2d, their northward deflection; and 3d, the exertion of an excessive amount of lake influence upon the east side. All this is illustrated by tracing the isotherm of 22°. Coming within the limits of the chart a few miles southwest of Omaha, it pursues an undulating course eastward to Ottawa, in Illinois, when it bends abruptly northward, passing west of Chicago, and east of Milwaukie, to Northport, at the mouth of Grand Traverse Bay, whence it bends southward to Corunna, in the middle of the lower peninsula of Michigan, and northward again to Thunder Bay Island of Lake Huron, and thence east to Penetanguishene on Georgian Bay. The isotherm of 23° reaches almost as far north; but, in crossing the peninsula of Michigan, it strikes southward into northern Indiana and Ohio, thence northward again almost to Thunder Bay Island. The sinuosities of this isotherm spread over a belt of four and one half degrees, or 300 miles in width. In other words, the influence of the lakes is such that the mean temperature of January at Northport and Thunder Bay Island is identical with that of Omaha, Peoria, Chicago and Fort Wayne. The January temperature of Mackinac and Marquette is the same as that of Green Bay and Fort Winnebago.

An island of cold is again indicated in the southeastern part of the peninsula of Michigan. In this case, its form and position correspond quite exactly with a region of elevation. The area in northern Iowa which, in July, is an island of warmth, appears to be in January, an island of cold. A similar one exists in the elevated region of southern Wisconsin, while a remarkable axis of cold stretches through northern Wisconsin and Minnesota. The axis is not entirely coincident with the crest of the ridge dividing the tributaries of Lake Superior from those of the Mississippi; since the warming influence of Lake Superior crowds it about 60 miles southward.

One of the most striking phenomena exhibited by the chart for January, is the excess of the warming influence along the eastern side of Lake Michigan. The isotherm of 23½° strikes from Chicago directly to Northport, almost at the opposite end of the lake. The contrast in January temperature between the opposite shores of the lake is, for the northern half, four degrees, and for the southern, six degrees. This circumstance is due to the fact that the prevailing winds of the region, during January, and indeed during the entire winter, come from the west and southwest, and are at the same time, the coldest winds. The precise ratios of all the winds from the east and the west of the meridian, in January, are, at Chicago, according to eleven years' observations, as 72 : 5=14.4; at Milwaukie, for thirteen years, as 60 : 18=3.33; at Manitowoc, for eleven years, as 67 : 11=6.09. These results embody all January winds, except those directly from the north or south. Thus the winds from the west of the meridian are greater in amount as well as severity. The reason why the excess of warming influence on the east side is greater toward the south than toward the north is evidently because north, and even northwest, winds coming from Green Bay, add their warming effect to that of Lake

Michigan, in all the region north of Milwaukie.

The isothermal charts for the summer and winter contrast in the same way as those for July and January, though the contrast is naturally less marked. From the summer chart we perceive that the isothermal of 72° makes its advent upon the northern limit of the chart, and disappears upon its southern limit, only 12° of longitude further east. Coming from the Winnipeg country, it passes near Dubuque and Ottawa, thence into the centre of the peninsula of Michigan. Sweeping around this region, it strikes directly south to Germantown and Portsmouth in Ohio. The summer temperature of the Winnipeg region and of central Michigan, is identical with that of northern Illinois and southern Ohio. Areas of cold exist in southern Michigan and northern Minnesota; and large areas of uniform temperature in Wisconsin, Indiana and Ohio.

The excess of cooling influence upon the west side of the lake, during the entire summer, is quite noticeable. The isothermals, in approaching the Lake Superior region, make an angle of 45° with the meridian; and, under the influence of Lake

Michigan, they become quite parallel with the meridian. It does not appear that, in the Lake Superior region, any excess of winds from that lake exists; but, in the vicinity of Lake Michigan, such excess is well established. At Chicago, the winds from the lake are to those from the land, during summer, as 151 : 119=1.27; at Milwaukie, the lake winds are to the land winds as 142 : 104=1.27; at Manitowoc, the lake winds are to the land winds as 153 : 123=1.24.

From the winter chart we notice that the isotherm of 24° undulates over a breadth of more than 200 miles. Other isotherms are similarly sinuated. The mean winter climate of Mackinac is 20°; and is identical with that of Green Bay, Fort Winnebago and Fort Dodge.

The excess of the warming influence on the east side of Lake Michigan is most apparent. The winter mean of Chicago is 24½°, while that of New Buffalo, in the same latitude, is 28°. The winter mean of Milwaukie is 22°, while that of its *vis-a-vis*, Grand Haven, is 26°. The winter mean of Fort Howard is 20° and of Appleton, 19°; while that of Traverse City, farther north than either, is 23½°. These contrasts illustrate again the

effect of the prevalence, during the cold season, of winds from the west of the meridian.

As to the isothermals for the spring and autumn, it might be expected that they would suffer little deflection under the influence of the lakes. Comparatively speaking, this is the case; but it will be noticed, nevertheless, that a marked cooling influence is exerted in the spring; since the isotherm of 43°, for instance, is deflected southward one hundred and fifty miles. It is worthy of remark, at the same time, that the maximum deflection takes place on the west side of Lake Michigan. On the east side, the deflection of the same isotherm amounts to no more than twenty miles. In general, we find the mean spring temperature of the eastern side of Lake Michigan to be about three degrees higher than the mean spring temperature of the western side. As this excess is accumulated in April and May,—especially in May—it is at once apparent that the circumstance has a most important bearing upon the growth of spring crops on the opposite sides of the lake. The effect is such that the temperature of Grand Haven, March 15, is equal to that of Milwaukie, March 21; that of Grand Haven, April 15, is equal to that of Milwaukie, April 24; that of Grand Haven, May 15, is equal to that of Milwaukie, May 28. These contrasts relate to *mean* temperatures. They show that vegetation on the east side secures a start of six to thirteen days. Add to this, protection from *exceptional* cold, in the form of spring frosts, and, to this, the effects of a drier and lighter soil, and we get a clear and demonstrative explanation of the difference of the agricultural and pomological products of the opposite sides of the lake.

This contrast of temperatures in spring is explained, as before, by the predominance, during the cold month of March, of winds from the west of the meridian, and, during the warmer months of April and May, of winds from the east of the meridian. Thus, at Manitowoc, in March, the winds from the west of the meridian are to those from the east, as 43 : 24=1.8; at Milwaukie, they are as 44 : 32=1.4; at Chicago, as 57 : 20=2.85. On the contrary, the preponderance of winds from the east of the meridian, during May, is, at Manitowoc, as 37 : 26=1.42; at Milwaukie, as 62 : 24=2.58; and in April, as 52 : 33=1.6. At Chicago, including north winds, which are here lake winds, the ratio of lake and land winds, in May, is as 44 : 40=1.1.

In autumn, the resultant of the lake influences on the west side is almost zero; while, on the east of Lake Michigan, a warming effect is experienced, amounting, along the southern half of the lake, to one or two degrees, and, along the northern half of the lake, to three or four degrees. This, as before, is caused by a preponderance, during each of the autumn months, of winds from the west of the meridian. This preponderance is shown, for Chicago, by the ratio of 151 : 70=2.16; for Milwaukie, by the ratio of 147 : 94=1.56, and for Manitowoc, by the ratio of 160 : 60=2.67.

The advantages thus secured to vegetation along the east side of the lake are not less in autumn than in spring. These singular facts depend upon a shifting of the prevalent winds at the end of the cold season, toward the close of March, and again at the end of the mild season near the close of November. An investigation of the monthly means on the opposite sides of the lake, during autumn, shows that the temperature attained at Milwaukie, October 15, is not reached at Grand Haven until October 20. The Milwaukie temperature of November 15 is only reached at Grand Haven, November 23. Comparing Chicago and New Buffalo, we find the Chicago temperature of September 15 is the same as the New Buffalo temperature of September 21. The October and November temperatures seem to be nearly coincident. These comparisons show that the warm season is lengthened on the east side, about six to eight days in the autumn. This, added to the time gained in the spring, makes the growing season, on the east side of Lake Michigan, from twelve to twenty-one days longer than on the west side—to say nothing about exemption from unseasonable frosts and a much warmer constitution of the soil upon the east side.

Turning our attention, now, to the chart of isothermals for the year, we might anticipate that the warming and cooling influences of the lakes would exactly neutralize each other, so that the isothermals would experience no deflection. We find, however, that on the western side the resultant influence is slightly cooling, and on the eastern side, decidedly warming. The resultant of these two influences gives a final resultant of a warming character exerted upon the eastern side.

This final resultant has a value of one-half to two degrees. In other words, Lake Michigan elevates the mean annual temperature of the contiguous region nearly two degrees above the norm. This results, of course, from the fact that the mean temperature of the lake waters is higher than that of the land. This excess must be considerably greater than the resultant warming influence upon the land. Its explanation is a curious and interesting subject of inquiry. It cannot be caused, as in the case of the Gulf Stream, by great currents moving from tropical regions. Nor can we attribute it to a large volume of river water poured into the lake from regions lying to the southward. Some more occult cause operates to raise the mean temperature of the lake above the normal temperature of the land. Some suggestions as to the nature of that cause have been offered by the writer on a former occasion, but it would be foreign to our purpose to introduce the discussion in this place.

In studying the influence of the great lakes upon the climate of the contiguous regions, we should especially note its presence under circumstances of exceptional cold or heat upon the land. For the purpose of illustrating these relations, we have constructed two isothermal charts for minimum temperatures. One of these is a chart for *mean minima*, and the other a chart for *extreme minima.* By the "mean minimum" of a locality, is meant the *average* of the yearly minima for a series of years; and by the "extreme minimum," the *lowest point* attained during that series of years. These charts present results which are truly striking. The isotherms in the vicinity of lakes Huron and Michigan, trend literally north and south. In the chart of mean minima, the isotherm of — 15° strikes from Mackinac through Manitowoc, Milwaukie, and New Buffalo, to Fort Riley, in Kansas, near the parallel of 39°. Here is a deflection over nearly seven degrees of latitude, or about 480 miles in a straight line. The meaning of this is, that the most excessive cold at Mackinac, for a period of 28 years, is not, on the average greater than at Fort Riley, 480 miles further south. It is one degree less than at Chicago for a term of eleven years. By a glance at the chart of extreme minima, we perceive, that the lowest point reached at Mackinac is but two degrees lower than the extreme minimum of St. Louis. Extreme weather of

Chicago is twelve degrees colder than at New Buffalo. The lowest extreme of Milwaukie is fourteen degrees below the extreme minimum of Grand Haven, while the extreme of Fort Howard is twenty degrees below that of Northport. In general, while the mean minimum along the west side of Lake Michigan is — 16°; that along the east side is —6° ; while the extreme minimum on the west side is — 22° to — 30°, that of the east side is — 10° to — 16°.

It is proper to direct attention to the important bearing of these additional facts upon the results of soil-cultivation. It will be remembered that it is not the severity of the winter mean, but that of the winter *extremes* which conditions the immunity of exotic plants from destructive frost. One killing freeze is as fatal as thirty. That one killing freeze is as likely to occur at Fort Riley, or Leavenworth, or Peoria, or even at St. Louis, as at Mackinac. The whole east shore of Lake Michigan is 15° to 20° more secure than any of the places just named. As grapes and peach trees require for their destruction, a temperature of — 20°, it is apparent that peach orchards and vineyards are perfectly secure along the whole extent of the eastern shore of Lake Michigan.*

The *rationale* of these climatic effects is not difficult to discover. It lies in the comparatively low capacity of watery surfaces for absorbing and radiating heat. The mean temperature of the land, in the middle latitude of Lake Michigan, is about 44½°, and that of the lake, a few degrees higher. In July, the temperature of the land rises to 74° while that of the lake is not above 51° or 52°. This difference is partly due to the fact that upon the land the heat from the solar rays is accumulated near the surface, while upon the water it is disseminated through the whole mass, at least to a considerable extent, by the action of waves and currents. In January, the mean temperatnre of the land sinks to 19°, while that of the water does not, probably fall below 40°. The atmosphere in contact with the water must partake, to some extent, of the temperature of the water, and, when moving from the water to the land, must

* This statement refers only to exemption from winter-killing. As the northern portions of this belt enjoy a smaller aggregate of summer warmth than the southern, another climatic condition enters to determine the actual productiveness of different portions of the belt, and to affect the comparison between the whole belt and regions further south in Illinois and Missouri.

transfer to the land, some portion of the heat or cold proper to the lake. The effect is a tendency to equalize the land temperatures in summer and winter. This tendency is most distinctly felt in case of extreme weather. On occasion of our coldest weather, the wind blows generally from the southwest, and, passing diagonally over lake Michigan for a distance of 100 to 200 miles must necessarily experience a great degree of amelioration.

In this connection, it is worth while to point out the fact that the arcuation of the longitudinal axis of Lake Michigan is such that a southwest wind striking the Grand Traverse region, must have passed over a much greater breadth of lake-surface than the same wind, in striking the region of St. Joseph; and hence the amelioration of winter extremes must be more marked in the former region than in the latter. It is further obvious that in the rare case of absolute calm or a southerly wind at a time of extreme cold, no portion of the peninsula would experience the warming influence of the lake.

The foregoing generalizations from the numerical data of the science of meteorology are abundantly confirmed by the results of the attempts made during a few years past to introduce the cultivation of peaches, grapes and other fruits along the entire belt from St. Joseph to Grand Traverse Bay. These results are so much a success that it is now generally acknowledged that scarcely a superior fruit-producing region exists within the United States.

The influence of the sea in equalizing temperatures has long been understood. The immunity from unseasonable frosts secured by bodies of fresh water to localities in their immediate vicinity, has also been universally observed; but the fact that inland lakes, of the size of Lake Michigan, exert an ameliorating agency quite comparable with that of the Atlantic Ocean, is something which has only been brought to light by recent thorough discussions of a wide range of meteorological data. On general principles, it has, indeed, been asserted by Professor Henry and by Blodget, and, at an earlier period, by Humboldt, that the great lakes of North America must exert some influence in deflecting the isothermal lines; but when we come to examine any of the charts which have been published to represent existing knowledge or conceptions, we fail to detect any marked inflection of these lines in passing the region of the great lakes. In fact, the thermometric observations from the fifty-five meteorological stations in Michigan have not heretofore been employed in tracing out the remarkable tortuosities of the isothermals of the lower peninsula of Michigan. These disclosures are destined to take their place among the most interesting phenomena of climatological science.*

We do not deem it expedient to extend this paper by the introduction of barometrical and psychrometrical results; but the distribution of rain and snow is a climatic element of such paramount economical importance that we think a summarized table may be acceptable. We have, accordingly, selected

* The foregoing general results were embodied in a popular paper published (with reduced isothermal charts for July and January) in Harper's Magazine for July, 1871. This paper, with the charts, has been reproduced in *Der Michigan Wegweiser*, in Hamburg, and also in the *Zeitschrift der œsterreichischen Gesellschaft für Meteorologie* in Vienna, Vol. viii. p. 40, *et seq*. (February 1, 1873.) It seems a suggestive commentary on the intelligence of American state governments that, while these results, though thus meagrely set forth, possess such interest as to be published and republished at home and abroad, by newspaper and magazine managers, emigration agencies, learned societies, medical journals and horticultural associations, the public authorities of Michigan have neither instigated, aided nor endorsed their publication ; but, incredible as it may seem, have actually declined, with expressions of derision, to publish them to their own citizens and the world. [See Michigan Legislative Proceedings, March and April, 1871.]

from our voluminous records of results the following condensed view of the aqueous precipitation at a series of representative localities.

The mean annual precipitation over the whole State is 31 inches; in the Upper Peninsula, 30 inches, and in the Lower, 32 inches. This is about the average for Wisconsin, Minnesota, Iowa, Nebraska and Kansas. In the states south and east of Michigan, the annual fall of rain and snow reaches 40 to 44 inches. Further south, and along the Atlantic border, it rises still higher. The total precipitation throughout the lake-region sustains no discoverable relation to the great lakes. Aside from the varying influence of the great current of moisture from the Gulf of Mexico, the precipitation seems to vary with the topography and surface of the country. It is singular, however, that, in this State, the four localities receiving the lowest mean annual precipitation, are situated upon the lake shores. These are Tawas, Ontonagon, Mackinac, and Grand Haven. On the contrary, however, two other localities, Copper Falls and Holland, situated in close proximity to the lakes, are exceeded only by Grand Rapids.

The mode of distribution of the precipitation through the year is a question which has an important bearing on the ability of a region to sustain an agricultural industry. The Table referred to has columns headed "Ratio," in which is placed, for each season, the ratio of the precipitation for that season to the whole annual precipitation at the same place. These ratios are expressed in the form of percentages. From these percentages we have calculated the following generalized Table.

DISTRIBUTION OF PRECIPITATION THROUGH THE SEASONS.

(In percentages of Total Precipitation.)

	SPRING.	SUMMER.	AUTUMN.	WINTER.
Upper Peninsula,	19	27	28.8	22
Lower Peninsula,	25.8	28.7	27.3	19.1
Whole State,	23.8	28.3	27.7	20

In the State at large, we have, as appears, considerably less precipitation during winter than during any other season. The Lower Peninsula presents this deficiency to a marked extent, while, in the Upper Peninsula, the spring is the period of minimum precipitation, though Copper Falls has a marked winter excess. In the whole State, and in the Lower Peninsula, the summer season is marked by the greatest amount of rain; in the Upper Peninsula, the autumn.

In the Lower Peninsula the three seasons of vegetable growth together receive nearly 82 per cent. of the whole precipitation. That is, the rain-fall, during the growing months, is as great as in other states having a total precipitation of 35 inches distributed equally through the seasons.

The liability of a region to occasional excessive droughts is not indicated by the total mean annual precipitation, nor, indeed, by the mean seasonal precipitation. An occasional prolonged and destructive period of dryness may occur without materially disturbing the annual or seasonal means. We have, accordingly, selected from the annual and seasonal means for a series of years, the ones which are lowest for each locality, and introduced them in our Table, in the columns headed "minima." From the column of minima for the year, we observe that the extreme minimum at Sault Ste. Marie is 12.11 inches, which is only 40 per cent. of the annual amount, and at Mackinac it is only 48 per cent. of the amount at that place. These numbers represent years of extreme scantiness of rain and snow. Had we the data, it would probably appear that the year 1871 was a year of remarkable dryness throughout the State. Generally, the extreme minimum of annual precipitation does not fall excessively below the normal annual mean. At Detroit, it is 60 per cent. of the annual mean; at Lansing, 81 per cent.; at Ann Arbor and Monroe, 82 per cent.; at Ontonagon, 83 per cent.; at Tawas, 84 per cent.; at Grand Haven, 87 per cent.; at Grand Rapids, 92 per cent., and at Marquette and Holland, 93 per cent., showing a remarkably uniform distribution through a series of years.

The extreme minima of the seasons exhibit a much greater departure from the normal seasonal means. For instance, in spring, the extreme minimum precipitation at Mackinac is only 33 per cent. of the norm; at Sault Ste. Marie, 34 per cent.; at Marquette, 43; at Ontonagon, 50; at Battle Creek and Ann Arbor, 54; at Detroit, 55; at Monroe, 56; at Grand Rapids, 60; at Thunder Bay I., 61; at Lansing, 81; at Holland, 86 per cent. Thus extreme dryness in spring is less severe in the lower peninsula than in the upper.

In summer, the extreme minimum precipitation at Mackinac is 34 per cent. of the norm; at Tawas, 38 per cent; at Sault Ste.

Marie, 39; at Detroit, 41; at Marquette, 45; at Battle Creek and Ann Arbor, 52; at Lansing and Grand Rapids, 58; at Monroe, 62; at Ontonagon, Holland and Flint, 68; at Grand Haven, 90 per cent. This means that the liability to extreme dryness throughout the summer is greater at Tawas, Sault Ste. Marie, Detroit and Marquette than at the other places; and that at Grand Haven, the normal supply is never diminished more than one tenth. The trustworthiness of these generalizations, however, is only in proportion to the length of the period of observations at the several places.

In autumn, the extreme minimum of precipitation at Mackinac is only 22 per cent. of the normal precipitation for that season; at Tawas, it is 38 per cent.; at Marquette, 39; at Lansing, 40; at Detroit, 43; at Monroe, 47; at Battle Creek, 51; at Sault Ste. Marie, 52; at Grand Haven, 53; at Grand Rapids 60; at Flint, 67; at Ann Arbor, 71; at Ontonagon, 75; and at Holland, 97 per cent. of the norm.

In winter, the extreme minimum precipitation at Detroit is 31 per cent. of the normal amount; at Mackinac, it is 38 per cent.; at Sault Ste. Marie, 49; at Monroe, 56; at

Ann Arbor, 60 ; at Battle Creek, 62 ; at Marquette, 64 ; at Ontonagon, 66 ; at Lansing, 69 ; at Tawas, 74 ; at Grand Haven, 76 ; at Holland, 90 per cent.

From the foregoing generalizations, it appears that the northern localities experience a somewhat greater liability to dryness in all the seasons. It must be borne in mind, however, that the percentages given are percentages of the seasonal means at the several localities. But this mean may be comparatively low. Thus, when we state extreme winter dryness at Ann Arbor as 60 per cent of the normal precipitation, it will be remembered that the normal precipitation, in winter, is only 15 per cent of the whole annual precipitation.

It is apparent that the seasonal minima are more excessive than the annual minima. It follows from this, that a deficiency of precipitation in one season is followed, within twelve months, by an excess in another season. This accords with popular belief.*

* The foregoing results are liable to be changed by further observations—the most so, at localities where the series of observations has not extended over a number of years. Of all the results, the extreme minima are most liable to undergo change. It will be noticed that the minima given are generally most extreme at those localities where the series of observations is most extended.

We append, finally, a condensed Table of the Winds of the State. The numbers in the columns denote the number of tri-daily observations, in each season and during the year, at which the wind, at the several localities, was from the directions indicated at the heads of the columns. Thus, at Ontonagon, in the spring, as the average result of three years' observations, the wind was found from the north 39 times, from the northeast, 45 times, and so on. The column headed "Ca" denotes the number of times a calm prevailed. Some observers report no calm—deciding always that there exists some determinable movement of the air, however slight. Hence the blanks in this column. Many interesting generalizations might be based upon the Table, some of which have already been presented in connection with the discussion of isothermals, but we forbear to extend this paper.

The foregoing popularized abstract of meteorological results is but a meagre exhibit of the amount of information in our possession, but the presentation is probably sufficiently full for the present purpose.

WINDS.

	SPRING.								
	N	NE	E	SE	S	SW	W	NW	Ca.
Ontonagon	39	45	9	0	65	10	26	32	46
Marquette	23	17	23	33	23	14	39	69	27
Sault Ste. Marie	28	18	52	57	23	26	46	57	
Mackinac	34	26	64	34	16	25	61	57	
St. James	19	23	33	19	16	84	16	52	14
Thunder Bay I.	47	32	28	68	34	19	11	57	29
Tawas City	4	40	22	15	22	35	40	56	
Ottawa Pt.	20	132	25	28	34	58	17	54	
Grand Haven	43	7	59	13	26	27	30	14	53
Flint	47	19	8	11	44	40	49	32	
Grand Rapids	14	44	31	20	16	55	35	24	33
Fort Gratiot	38	74	16	29	36	60	28	34	
Port Huron	57	16	1	21	28	44	26	44	10
Lansing	14	36	27	20	13	61	42	17	20
Cooper	8	1	27	1	3	2	157	5	68
Battle Creek	14	19	33	27	28	44	53	31	22
Detroit	29	58	42	21	30	56	45	25	
Ann Arbor	24	30	20	30	17	46	42	40	
Monroe	11	43	27	33	13	39	28	76	

WINDS.

	SUMMER.								
	N	NE	E	SE	S	SW	W	NW	Ca.
Ontonagon	41	37	10	9	54	14	30	29	28
Marquette	23	14	27	26	25	28	43	51	21
Sault Ste. Marie	19	15	33	50	28	81	42	71	
Mackinac	33	21	49	29	22	25	87	49	
St. James	32	7	25	20	16	107	6	44	24
Thunder Bay I.	62	20	11	52	56	26	14	50	24
Tawas City	14	24	24	13	41	40	43	37	
Ottawa Pt.	20	67	34	42	66	56	18	64	
Grand Haven	23	11	33	10	28	34	59	17	33
Flint	37	9	6	6	52	41	32	27	
Grand Rapids	9	33	17	17	9	73	43	28	46
Fort Gratiot	31	73	12	34	44	42	17	30	
Port Huron	62	10	0	16	41	77	11	57	4
Lansing	12	28	19	15	14	78	40	14	55
Cooper	1	1	24	3	3	3	82	5	140
Battle Creek	15	16	18	14	42	47	42	22	53
Detroit	25	40	35	25	47	56	47	26	
Ann Arbor	16	19	18	26	27	47	28	33	
Monroe	14	42	15	41	14	58	16	69	

WINDS.

	AUTUMN.								
	N	NE	E	SE	S	SW	W	NW	Ca.
Ontonagon	23	33	5	16	68	20	35	30	20
Marquette	20	18	19	21	32	31	43	54	18
Sault Ste. Marie	33	26	45	48	31	32	35	50	
Mackinac	37	21	41	29	29	33	64	65	
St. James	23	15	30	26	20	73	5	68	14
Thunder Bay I.	34	24	23	45	50	35	27	64	9
Tawas City	9	18	10	12	32	45	40	36	
Ottawa Pt.	25	47	19	24	37	76	35	80	
Grand Haven	17	30	42	23	23	44	46	28	15
Flint	43	12	4	8	59	43	28	26	
Grand Rapids	13	37	22	17	18	65	36	22	43
Fort Gratiot	19	44	10	35	41	80	30	41	
Port Huron	27	17	2	13	27	88	26	33	14
Lansing	10	35	18	13	14	85	35	26	48
Cooper	2	2	23	2	5	4	111	6	117
Battle Creek	11	16	23	19	50	45	48	27	34
Detroit	26	28	29	17	38	57	62	30	
Ann Arbor	20	17	17	35	37	66	37	33	
Monroe	20	31	15	30	18	57	27	65	

WINDS.

	WINTER.								
	N	NE	E	SE	S	SW	W	NW	Ca.
Ontonagon	36	12	4	2	81	10	41	39	39
Marquette	19	9	16	27	28	22	69	53	15
Sault Ste. Marie	30	27	57	52	28	31	30	38	
Mackinac	41	34	27	32	29	27	63	56	
St. James	19	20	13	20	13	60	38	72	11
Thunder Bay I.	40	15	12	30	46	39	28	77	15
Tawas City	15	13	9	12	30	52	45	48	
Ottawa Pt.	15	37	11	35	56	92	27	88	
Grand Haven	32	19	35	13	23	36	58	24	30
Flint	35	10	7	32	51	39	43	33	
Grand Rapids	11	33	22	23	18	62	46	25	28
Fort Gratiot	21	32	11	33	42	87	39	47	
Port Huron	7	16	6	25	42	76	40	44	6
Lansing	7	16	17	29	11	99	50	24	23
Cooper	7	4	35	2	8	20	133	7	53
Battle Creek	9	12	33	22	26	50	65	25	20
Detroit	27	32	23	21	34	70	58	33	
Ann Arbor	18	19	17	24	19	68	37	43	
Monroe	15	31	13	23	16	58	34	71	

WINDS.

	YEAR.								
	N	NE	E	SE	S	SW	W	NW	Ca.
Ontonagon	139	127	28	27	268	54	132	130	133
Marquette	85	58	85	107	108	95	194	227	81
Sault Ste. Marie	110	86	187	201	110	120	153	217	
Mackinac	145	102	183	124	96	110	275	227	
St. James	93	65	101	85	65	324	65	236	63
Thunder Bay I.	183	91	74	195	186	119	80	242	77
Tawas City	42	95	65	52	125	172	208	177	
Ottawa Pt.	80	283	89	129	193	282	87	286	
Grand Haven	115	67	169	59	100	144	193	83	131
Flint	162	50	25	57	206	163	152	118	
Grand Rapids	47	147	92	77	61	255	160	99	150
Fort Gratiot	109	223	49	131	163	259	114	152	
Port Huron	153	59	9	75	138	285	103	178	34
Lansing	43	115	81	77	52	325	167	75	146
Cooper	18	8	109	8	19	29	483	23	378
Battle Creek	49	63	107	82	146	186	210	105	129*
Detroit	107	158	129	84	149	239	212	114	
Ann Arbor	78	85	72	115	100	227	144	149	
Monroe	58	147	70	127	61	212	10 5	281	

Precipitation of Rain and Snow.

LOCALITY.	Latitude.	Elev. above sea. Feet.	Number of years.	SPRING.			SUMMER.			AUTUMN.			WINTER.			YEAR.	
				MEAN.		MIN.	MEAN.		MIN.	MEAN.		MIN.	MEAN.		MIN.	MEAN.	MIN.
				inches	ratio	Inches.	inches	ratio	Inches.	inches	ratio	Inches.	inches	ratio	Inches.	Inches	Inches
Copper Falls	47° 25′	1200	5	7.25	19	6.08	7.23	19	5.15	9.83	26.4	6.78	12.92	34.7	11.79	37.23	30.09
Ontonagon	46° 52′	630	12	4.71	19	2.34	7·21	30	4.88	6.25	25.8	4.68	6.02	24.9	3.98	24.20	20.09
Marquette	46° 32′	625	13	7.13	23	3.08	8.90	29	3.99	8.85	28.5	3.44	6.14	10.7	3.95	31.02	28.84
Sault Ste. Marie	46° 30′	610	33	5.07	15.7	1.74	9.43	31	3.69	10.46	34.5	5.45	5.03	16.6	2.46	30.28	12.11
Mackinac	45° 51′	731	28	4.59	19	1.53	9.08	37	3.11	7.06	28.7	1.58	3.58	14.5	1.35	24.58	11.70
Northport	45° 03′	600	4				11.96			14.68			6.83				
Thunder Bay I.	45° 02′	590	7	7.97	23.7	4.85	7.72	23	6.10	9.28	27 6	6.51	8.84	26.3	6.17	33.61	27.88
Tawas City	44° 15′	583	11	4.67	22	2.83	6.01	29	2.31	6.41	30.5	2.44	3.89	18	2.88	20.99	17.69
Grand Haven	43° 03′	588	4	5.33	21	3.94	8.59	34	7.77	7.72	30.4	4.13	4.64	18.3	3.54	25.32	21.93
Flint	43° 01′	400	3			6.78	9.41		6.40	10.40		6.92			7.48		40.55
Grand Rapids	42° 58′	360	11	11.37	28.6	6.79	10.08	25.3	5.90	9.73	24.4	5.82	8.62	22		39.81	36.73
Fort Gratiot	43° 00′	598	18	8.02	24.5	6.06	9.75	30	6.56	8.86	27	4.93	5.75	17.6	4.35	32.62	25.75
Holland	42° 42′	595	4	9.37	24	8.10	7.31	19	4.97	12.21	31 5	11.92	9.91	25.5	8.97	38.80	35.99
Lansing	42° 36′	850	7	8.24	27	6.68	9.57	31.5	5.52	7.06	23.3	2.80	5.52	18.2	3.84	30.31	24.58
Battle Creek	42° 16′	510	6½	9.12	29	4.96	7.88	25	4 10	8.94	28.5	4.54	5.45	17.3	3.36	31.39	25.73
Detroit	42° 19′	595	30	8.60	24.5	4.80	11.15	31.7	4.56	9.28	26.4	4.03	6.05	17.2	1.88	35.09	21.10
Ann Arbor	42° 16′	868	7	8.14	25	4.43	11.05	34.5	5.77	7.97	25	5.69	4.82	15	3.00	31.98	26.26
Litchfield*	41° 58′	1007	3½	13 24	29	11.74	12.69	28	9.41	12.21	26.8	8.44	7.37	16	6.17	45.57	37.59
Monroe	41° 53′	584	18	8.11	25.5	4.56	9.85	31	6.10	8.27	26	3.87	5.56	17.4	3.11	31.80	26.17

* The amount of precipitation reported for this place is so abnormally large that the results have been employed only in calculating the seasonal percentages.

This Table is based on observations extended, generally, to the year 1870, inclusive. Since the results were worked out, the volume of *Tables and Results of the Precipitation in Rain and Snow, in the United States*, compiled and discussed by Charles A. Schott of the U. S. Coast Survey, has been published by the Smithsonian Institution. Some slight discrepancies with our determinations may, undoubtedly, be attributed to the fact that Mr. Schott's data do not, generally, extend beyond 1866 or 1867. This Report, like everything which emanates from Mr. Schott's hands, is a masterly work. The country is deeply indebted to the Smithsonian Institution for meteorological data and discussions unsurpassed in volume and value by the productions of any country.

www.ingramcontent.com/pod-product-compliance
Lightning Source LLC
LaVergne TN
LVHW021423110826
845150LV00007B/2058

* 9 7 8 1 4 2 5 5 0 8 3 1 9 *